Lecture Notes in Mathematics

Edited by A. Dold and B. Eckmann

506

Numerical Analysis

Proceedings of the Dundee Conference
on Numerical Analysis, 1975

Edited by G. A. Watson

Springer-Verlag
Berlin · Heidelberg · New York 1976

Editor

G. Alistair Watson
Department of Mathematics
University of Dundee
Dundee, Scotland

Library of Congress Cataloging in Publication Data

Dundee Conference on Numerical Analysis, 1975.
  Numerical analysis.

  (Lecture notes in mathematics ; 506)
  Conference held July 1-4, 1975, at the University of
Dundee.
  Bibliography: p.
  Includes index.
  1. Numerical analysis--Congresses. I. Watson,
G. A. II. Title. III. Series: Lecture notes in
mathematics (Berlin) ; 506.
QA3.L28 no. 506 [QA297] 510'.8s [519.4]   75-45241

AMS Subject Classifications (1970): 65-02, 65 D 05, 65 D 20, 65 D 30, 65 F 05, 65 F 20, 65 K 05, 65 L 05, 65 N 30, 65 P 05, 65 R 05, 90 C 05, 90 C 30

ISBN 3-540-07610-7 Springer-Verlag Berlin · Heidelberg · New York
ISBN 0-387-07610-7 Springer-Verlag New York · Heidelberg · Berlin

© by Springer-Verlag Berlin · Heidelberg 1976
Printed in Germany
Printing and binding: Beltz, Offsetdruck, 6944 Hemsbach/Bergstr.

Sep

1444119

Foreword

For the 4 days July 1-4, 1975, around 200 people attended the 6th biennial conference on numerical analysis at the University of Dundee, Scotland. Previous conferences have in the main been concerned with specific subject areas, such as the numerical solution of differential equations. This year, however, it was decided to broaden the scope of the meeting to encompass the whole of numerical analysis, while maintaining a bias towards the more practical side of the subject.

Invitations to present talks were accepted by 16 eminent numerical analysts, representative of a variety of fields of activity, and their papers appear in these notes. In addition to the invited papers, short contributions were solicited, and 45 of these were presented at the conference in parallel sessions. A list of these papers is given, together with the addresses of the authors (correct at the time of the conference). I would like to thank all speakers, including the after dinner speaker at the conference dinner, Mr A R Curtis, all chairmen and participants for their contributions.

It is not always realised that the Dundee numerical analysis conferences are firstly, financially self-supporting, and secondly, organised entirely from within the Department of Mathematics. As on so many previous occasions, the organisation of the conference was in the very capable hands of Dr J Ll Morris, assisted by various other members of the Mathematics Department. Particularly in view of the recent departure of Dr Morris from Dundee to the University of Waterloo, I would like to take this opportunity to pay tribute to the very considerable contribution he has made to the continued success of the numerical analysis conferences in Dundee.

The typing of the various documents associated with the conference and some of the typing in this volume has been done by secretaries in the Mathematics Department, in particular Miss R Dudgeon; this work is gratefully acknowledged.

G A Watson

Dundee, September 1975.

CONTENTS

INVITED SPEAKERS

R E Barnhill	Mathematics Department, University of Utah, Salt Lake City, Utah 84112, U.S.A.
H Brunner	Institut für Numerische Mathematik, Westfälische Wilhelms-Universität, D-44 Münster, Germany.
J C Butcher	Mathematics Department, The University of Auckland, Auckland, New Zealand.
W J Cody	Argonne National Laboratory, 9700 South Cass Avenue, Argonne, Illinois 60439, U.S.A.
L Collatz	Institut für Angewandte Mathematik, Universitat Hamburg, 2 Hamburg 13, Rothembaumchausse 67/69, West Germany.
A R Curtis	Computer Science and Systems Division, A.E.R.E. Harwell, Didcot, Oxfordshire, England.
G Dahlquist	Department of Computer Science, Royal Institute of Technology, S-100 44 Stockholm 70, Sweden.
R Fletcher	Mathematics Department, University of Dundee, Dundee DD1 4HN, Scotland.
W Gautschi	Computer Sciences Department, Purdue University, West Lafayette, Indiana 47907, U.S.A.
W M Gentleman	Computer Science Department, University of Waterloo, Waterloo, Ontario, Canada.
J W Jerome	Department of Mathematics, Northwestern University, Evanston, Illinois 60201, U.S.A.
C L Lawson	Jet Propulsion Laboratory, Californian Institute of Technology, 4800 Oak Grove Drive, Pasadena, California 91103, U.S.A.
O L Mangasarian	Computer Sciences Department, University of Wisconsin, 1210 West Dayton Street, Madison, Wisconsin 53706, U.S.A.
W Murray	Division of Numerical Analysis and Computing, National Physical Laboratory, Teddington, Middlesex TW11 OLW, England.
M J D Powell	Computer Science and Systems Division, A.E.R.E. Harwell, Didcot, Oxfordshire, England.
J K Reid	Computer Science and Systems Division, A.E.R.E. Harwell, Didcot, Oxfordshire, England.
H J Stetter	Institut für Numerische Mathematik, Technische Hochschule Wien, A-1040 Wien, Gusshausstr, 27-29 Austria.

Submitted Papers

A Z Aktas and H Öncül: Computer Science Dept., Middle East Technical University, Ankara, Turkey.
Some numerical methods for nonlinear boundary value problems in O.D.E's.

R Alt: Institut de Programmation, Faculte des Science, Universite de Paris, Tour 55 - 11 Quai Saint-Bernard, Paris 5.
Evaluation of the numerical error committed in the floating point computation of a scalar product.

D E Amos: Numerical Division, Sandia Laboratories, Albuquerque, New Mexico.
Computation of I and J Bessel functions for real, non-negative orders and arguments.

E Ball* and R A Sack+: *Dept of Electrical Engineering and +Dept of Mathematics, University of Salford, Salford, England.
Numerical quadrature of line integrals.

K E Barrett: Mathematics Dept, Lanchester Polytechnic, Coventry, England.
Applications and extension of a variational principle for the stream function-vorticity formulation of the Navier-Stokes equations incorporating no slip conditions.

C Brezinski: University of Lille, France.
Computation of Padé approximants.

C Carter: Trent University, Peterborough, Ontario, Canada.
Evaluation of the greatest eigenvalue of an irreducible non-negative matrix.

F H Chipman: Mathematics Dept, Acadia University, Wolfville, N.S., Canada.
Implicit A-stable R-K methods with parameters.

M G Cox: Division of Numerical Analysis and Computing, National Physical Laboratory, Teddington, Middlesex, England.
The numerical evaluation of a spline from its B-spline representation.

L M Delves and J M Watt: Department of Computational and Statistical Science, University of Liverpool, England.
A proposal for a Gauss quadrature library package.

J C Eilbeck and G R McGuire: Mathematics Dept, Heriot-Watt University, Riccarton, Currie, Midlothian, Scotland.
Finite difference methods for the solution of the regularized longwave equation.

N T S Evans* and A R Gourlay+: *MRC Cyclotron Unit, Hammersmith Hospital, London, England and +IBM UK Scientific Centre, Peterlee, Co Durham, England.
The solution of a diffusion problem concerned with oxygen metabolism in tissues.

R Fletcher and T L Freeman: Mathematics Dept, The University, Dundee, Scotland.
A modified Newton method for minimization.

W Forster: Mathematics Dept, The University, Southampton, England.
The structure of computational methods: A note on consistency, convergence, and stability.

R Frank: Institut für Numerische Mathematik, Technische Hochschule Wien, A-1040 Wien, Gusshausstr, 27-29 Austria.
The method of Iterated Defect-Correction.

T L Freeman, D F Griffiths and A R Mitchell: Mathematics Dept, The University, Dundee, Scotland.
Complementary variational principles and the finite element method.

J H Freilich and E L Ortiz: Mathematics Dept, Imperial College, London University, England.
Tau method approximation to the solution of 2nd order linear differential equations.

I Gargantini: University of Western Ontario, London, Ontario, Canada.
Parallel Laguerre iterations: The complex case.

E Hairer: Mathematics Dept, Université de Genève, Switzerland.
Equations of condition for Nystroem methods.

P J Hartley: Mathematics Dept, Lanchester Polytechnic, Priory Street, Coventry CV1 5FB, England.
Some tensor product, hypersurface fitting methods.

J G Hayes: National Physical Laboratory, Teddington, England.
Bicubic splines with curved knot-lines.

T R Hopkins* and R Wait+: *Computing Laboratory, University of Kent, Canterbury, England and +Dept of Computational and Statistical Science, University of Liverpool, Liverpool, England.
A comparison of numerical methods for the solution of quasi-linear P.D.E's.

E S Jones: Dept of Computing Science, The University of Glasgow, Glasgow, Scotland.
Quasi-Newton methods for non-linear equations: Line search criteria and a new update.

R B Kelman and J T Simpson: Dept of Computer Science, Colorado State University, Ft Collins, Colorado 80523, U.S.A.
Algorithms for solving dual trigonometric series.

F M Larkin: Dept of Computing and Information Science, Queen's University, Kingston, Ontario, Canada.
A note on the stability of Ritz-type discretizations of certain parabolic equations.

T Lyche: Mathematics Dept, University of Oslo, Oslo 3, Norway.
Asymptotic expansions and error bounds for cubic smoothing splines.

D Meek: Mathematics Dept, Brunel University, Uxbridge, Middlesex, England.
Toeplitz matrices with positive inverses.

C A Micchelli and A Pinkus: IBM, Research Division, Yorktown Heights, NY 10598, U.S.A.
On n-widths in L^{∞}.

H D Mittlemann: Fachbereich Mathematik, Der Technischen Hochschule Darmstadt, 61 Darmstadt, Kantplatz 1, Germany.
On pointwise estimates for a finite element solution of nonlinear boundary value problems.

E Moore: Memorial University of Newfoundland, St John's, Newfoundland, Canada.
Curve fitting using integral equations.

M Neumann: Israel Institute of Technology, Haifa, Israel.
Subproper splitting for rectangular matrices.

M A Noor: Mathematics Dept, Brunel University, Uxbridge, England.
Error bounds for the approximation of variational inequalities.

J Oliver: Computer Centre, University of Essex, Colchester, England.
A curiosity of low-order explicit Runge-Kutta methods.

I E Over, Jr: Lowell University, Lowell, Mass., U.S.A.
A modern course for training student engineers.

P D Panagiotopoulos and B R Witherden: Inst fur Technische Mechanik, R.W.T.H., 51 Aachen, Templegraben 64, W Germany.
On a system of hyperbolic variational inequalities.

J D Pryce* and B Hargrave+: *Dept of Computer Science, University of Bristol, Bristol, England and +Mathematics Dept, University of Aberdeen, Aberdeen, Scotland.
On the numerical solution of multiparameter eigenvalue problems in ordinary differential equations.

R Rautmann: Mathematics Dept, University of Paderborn, W Germany.
On Galerkin methods for stabilized Navier Stokes problems.

E Spedicato: CISE, PO Box 3986, 20100 Milan, Italy.
A three parameter class of quasi-Newton algorithms derived from invariancy to non-linear scaling.

P Spellucci: Mathematics Dept, University of Mainz, Germany.
A modification of Wittmeyers method.

W J Stewart: Laboratoire d'Informatique, Université de Rennes, France 35000.
Markov modelling using simultaneous iteration.

P G Thomsen and Z Zlatev: Institute for Numerical Analysis, Technical University of Denmark.
A two-parameter family of PECE methods and their application in a variable order, variable stepsize package.

G Varga: Computer and Automation Institute, Hungarian Academy of Sciences, Budapest, Hungary.
A relaxation method for computation of the generalized inverse of matrices.

M van Veldhuizen: Wiskundig Seminarium, Vrije Universiteit, Amsterdam, Netherlands.
A projection method for a singular problem.

J G Verwer: Mathematisch Centrum, Amsterdam, Netherlands.
S-stability for generalized Runge-Kutta methods.

I Zang and J-P Vial: Center for Operations Research and Econometrics, de Croylaan 54, 3030 Heverlee, Belgium.
Unconstrained optimization by approximation of the gradient path.

NONCONFORMING FINITE ELEMENTS FOR CURVED REGIONS

Robert E Barnhill and James H Brown

Introduction

Finite element analysis is the piecewise approximation of the solution of a problem in variational form. The variational principles frequently are associated with an elliptic boundary value problem. Certain approximations to the variational principle are used in practice. Strang calls three of these "variational crimes". They are described below. This paper considers solutions to each crime.

The form of the variational problem requires the approximations to have a certain smoothness in order that they "conform" to the theory, e.g. for fourth order problems, conforming elements have C^1 continuity. This requirement leads to complicated finite elements. Engineers have sometimes chosen simpler elements that are "nonconforming" in that they are not smooth enough to fit the theory. For example, over a triangulation, C^1 polynomial elements are of fifth degree whereas a frequently used element is of second degree. Some nonconforming elements converge in practice. Irons' "patch test" attempted to justify these results and has since been shown to be theoretically sound [10,5].

Two other aspects of the variational formulation are frequently approximated: the boundary conditions and the integrals involved in the finite element method. A general theory of interpolation to boundary conditions has recently been given by Barnhill and Gregory [2]. This method is combined with that of nonconforming elements to produce elements that interpolate all the boundary data exactly and pass the patch test when used with standard nonconforming elements. The topic of numerical integration is considered for the case of a curved triangle in which the curved side is a hyperbola.

1. Nonconforming elements and the patch test

In this section we suppose that we are given a linear, constant-coefficient, self-adjoint, elliptic differential operator A and that we are required to solve the associated p.d.e. : find u such that

$$\begin{aligned} Au &= f \quad \text{in} \quad R \subset \mathbb{R}^2 \\ Bu &= g \quad \text{on} \quad \partial R \end{aligned} \tag{1.1}$$

where B is a vector of boundary differential operators. The Ritz-Galerkin formulation of (1.1) is: find $u \in V$ such that

$$a(u,v) = (f,v) \qquad \forall v \in \overset{o}{V} \tag{1.2}$$

where $a(\cdot,\cdot)$ is the bilinear form associated with A and V is the space of admissible functions, $\overset{o}{V}$ being those with zero boundary values.

In the finite element method the region R is divided up into elements and consequently the discrete analogue of (1.2) is: find $u_h \in V_h$ such that

$$a_h(u_h,v_h) = (f,v_h)_h \qquad \forall v_h \in \overset{o}{V}_h \tag{1.3}$$

where V_h (resp. $\overset{o}{V}_h$) is a finite-dimensional subset of V(resp. $\overset{o}{V}$). The essential difference between (1.2) and (1.3) is that in (1.3) the energy $a(\cdot,\cdot)$ is calculated separately in each element whereas in (1.2) it is calculated over the region as a whole.

In the case of conforming elements (1.2) and (1.3) are equivalent, but in the case of nonconforming elements we do not have the inclusion $V_h \subset V$ and thus, although (1.3) is still well-defined, (1.3) and (1.2) are by no means equivalent since the energy in (1.2) becomes infinite. Since (1.3) makes sense, we can utilise (1.3) to find a nonconforming solution u_h ; this solution may not, however, converge to the true solution u . The behaviour of u_h depends crucially on an idea due to Irons [6]; this is now known as the <u>patch test</u>.

Suppose we have the following:

(i) the energy $a(\cdot,\cdot)$ contains derivatives of order k.

(ii) the nonconforming trial space is such that $\mathcal{P}_k \subset V_h$ (where $\mathcal{P}_k \equiv \{p_k : p_k$ is a polynomial of degree $\leq k\}$).

(iii) boundary conditions and the right hand side of the original p.d.e. are chosen so that the solution $u \equiv p_k \in \mathcal{P}_k$.

The patch test, as stated by Irons, then requires that the finite element solution $u_h \in V_h$ (calculated by solving (1.3) i.e. by ignoring inter-element discontinuities) is identically p_k .

The patch test was first examined from a mathematical viewpoint by Strang [10]. In a similar vein to Strang, we have the following: the patch test is passed if and only if

$$a_h(p_k,v_h) = (Ap_k,v_h)_h \quad . \tag{1.4}$$

A neat way of showing the equivalence of Irons' statement of the patch test with (1.4) is via the following inequalities (see Strang and Fix [11]). Define

$$\|u-u_h\| \equiv a_h(u-u_h,u-u_h)^{\frac{1}{2}} .$$

Then (i) $$\|u-u_h\| \geq \frac{|a_h(u,v_h) - (Au,v_h)_h|}{\|v_h\|} \qquad \forall v_h \in V_h \tag{1.5}$$

(ii) $$\|u-u_h\| \leq \max_{v_h \in V_h} \frac{|a_h(u,v_h) - (Au,v_h)_h|}{\|v_h\|} + \min_{v_h \in V_h} \|u-v_h\| \tag{1.6}$$

Irons $\Rightarrow$ (1.4): take the solution u to be $p_k \in \mathcal{P}_k$ i.e. $u \equiv p_k$. Then, according to Irons, if the patch test is passed, the finite element approximation $u_h \equiv p_k$. (1.5) now gives

$$0 = \|u-u_h\| \geq \frac{|a_h(p_k,v_h) - (Ap_k,v_h)_h|}{\|v_h\|}$$

and hence

$$a_h(p_k,v_h) = (Ap_k,v_h)_h$$

(1.4) $\Rightarrow$ Irons: suppose again that the solution $u \equiv p_k \in \mathcal{P}_k$ and that (1.4) holds. Then (1.6) gives

$$\|p_k-u_h\| \leq \min_{v_h \in V_h} \|p_k-v_h\|$$

But the right hand side is zero since $\mathcal{P}_k \subset V_h$. Thus $\|p_k-u_h\| = 0$, i.e. $u_h = p_k$. Thus (1.4) is equivalent to Irons' original statement of the patch test. The inequality (1.6), along with (1.4), is fundamental in obtaining convergence results for non-conforming trial spaces.

Since the functions v_h are smooth in each element, we may use integration by parts in (1.4) to obtain a further restatement of the patch test. In the case of second order problems, this procedure yields the following conditions which are sufficient for the patch test to be passed:

(a) on boundary segments ∂R_i,

$$\int_{\partial R_i} v_h d\sigma = 0 \tag{1.7a}$$

(b) on internal element sides ∂T_i,

$$\int_{\partial T_i} (v_h^I - v_h^{II}) d\sigma = 0 \tag{1.7b}$$

where v_h^I and v_h^{II} represent v_h in the elements sharing the side ∂T_i.

An example of a trial space which has properties (1.7a) and (1.7b) is the space of functions defined by:

(i) v_h is a polynomial of degree 1 in each triangle T.

(ii) v_h is continuous at the mid-points of the sides of T.

This element is denoted diagrammatically by:

Figure 1.1: Nonconforming linear element.

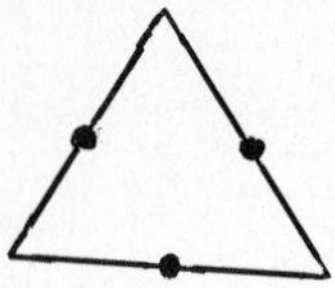

The following notation is used in the Figures:

• means interpolation to function value at a point.

┼ means interpolation to normal derivative at mid-point.

In the case of fourth order problems, <u>sufficient</u> conditions for the patch test to be passed are:

(a) on boundary segments ∂R_i,

$$\int_{\partial R_i} \frac{\partial v_h}{\partial n}\, d\sigma = 0 \quad \text{and} \quad \int_{\partial R_i} \frac{\partial v_h}{\partial t}\, d\sigma = 0 \tag{1.8a}$$

(b) on internal element sides ∂T_i,

$$\int_{\partial T_i} \left[\frac{\partial v_h^I}{\partial n} - \frac{\partial v_h^{II}}{\partial n}\right] d\sigma = 0 \quad \text{and} \quad \int_{\partial T_i} \left[\frac{\partial v_h^I}{\partial t} - \frac{\partial v_h^{II}}{\partial t}\right] d\sigma = 0 \tag{1.8b}$$

where $\partial/\partial n$ and $\partial/\partial t$ represent normal and tangential derivatives respectively.

An example of a trial space which has properties (1.8a) and (1.8b) is the space of functions defined by:

(i) v_h is a polynomial of degree 2 in each triangle T

(ii) v_h is continuous at the vertices of T

(iii) $\frac{\partial v_h}{\partial n}$ is continuous at the mid-points of the sides of T.

This element is the Morley triangle [8,9]. It is denoted diagrammatically by:

Figure 1.2: Morley triangle.

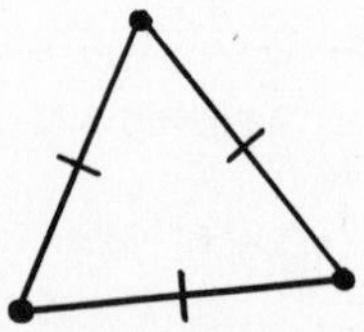

Ideas similar to those described above which also produce alternative patch test representations may be found in Ciarlet [4,5] and de Veubeke [12]. The line integral representations allow a simple a priori determination of whether or not a non-conforming trial space passes the patch test. More examples along with different patch test sufficient conditions may be found in Brown [3].

2. Boundary Conditions

If the boundary data in (1.1) are not interpolated exactly, then the error made constitutes a second variational crime. That is, the trial functions v_h do not belong to a subset of the test functions v in (1.2) and so the standard finite element analysis does not apply. In fact, numerical experiments have been done at Dundee in which the potential energy of a trial function was less than the minimum energy attained by the solution u of (1.2); this apparent contradiction being explained by the trial function not satisfying u's boundary conditions.

Barnhill and Gregory [2] have devised interpolation schemes that interpolate to function values and normal derivatives all around the boundary of a triangle with one curved side and two straight sides. There is a variety of interpolants to do this, but the ones presented here have polynomial weighting ("blending") functions whereas the competing interpolants have rational blending functions.

These interpolants can be used either as direct methods on a physical curved triangle in the x-y plane or as a mapping to a standard triangle, but we do not pursue this idea in this paper.

We consider the physical triangle to be given as shown in Figure 2.1(b), the curved side being given in the parametric form $x = x(t)$, $y = y(t)$. We transform the

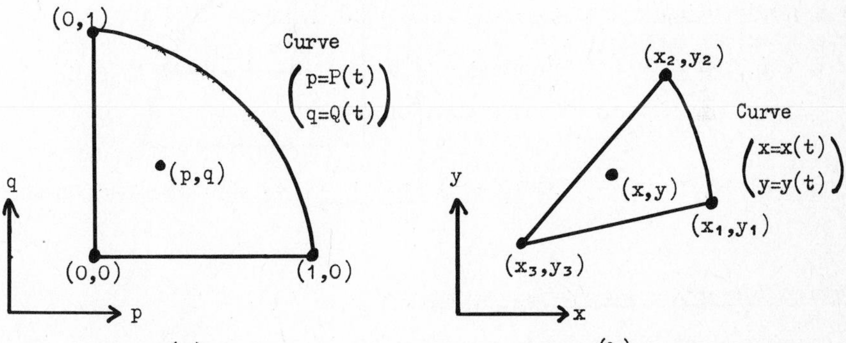

(a) Standard triangle in the p-q plane.

(b) Physical triangle in the x-y plane.

Figure 2.1

interpolation data from the physical triangle to the standard curved triangle shown in Figure 2.1(a), by using the following affine transformations:

$$x = px_1 + qx_2 + (1-p-q)x_3$$
$$y = py_1 + qy_2 + (1-p-q)y_3 \tag{2.1}$$

$$p = \frac{(x-x_3)(y_2-y_3) - (y-y_3)(x_2-x_3)}{(x_1-x_3)(y_2-y_3) - (y_1-y_3)(x_2-x_3)}$$
$$q = \frac{(y-y_3)(x_1-x_3) - (x-x_3)(y_1-y_3)}{(y_2-y_3)(x_1-x_3) - (x_2-x_3)(y_1-y_3)} \tag{2.2}$$

The given curve $\begin{pmatrix} x(t) \\ y(t) \end{pmatrix}$ in the physical plane is affinely transformed to the curve $\begin{pmatrix} P(t) \\ Q(t) \end{pmatrix}$ by means of (2.2) by setting $P(t) = p(x,y)\big|_{x=x(t),y=y(t)}$, $Q(t) = q(x,y)\big|_{x=x(t),y=y(t)}$. The given values of the function $F = F(x,y)$ are similarly transformed to $G = G(p,q) = F(x,y)\big|_{x=x(p,q),y=y(p,q)}$ by means of (2.1). The Barnhill-Gregory interpolant to function values all around the boundary of the triangle in Figure 2.2 is:

$$\begin{aligned} U(p,q) &= qG(p,Q(t_p)) + [1 - Q(t_p)]G(P(t_q),q) \\ &+ Q(t_p)G(0,q) + (1-q)G(p,0) + [q-Q(t_p)]G(0,0) \\ &- qG(0,Q(t_p)) - [1-Q(t_p)]G(P(t_q),0) \end{aligned} \tag{2.3}$$

where $t_p = P^{-1}(p)$, $t_q = Q^{-1}(q)$.

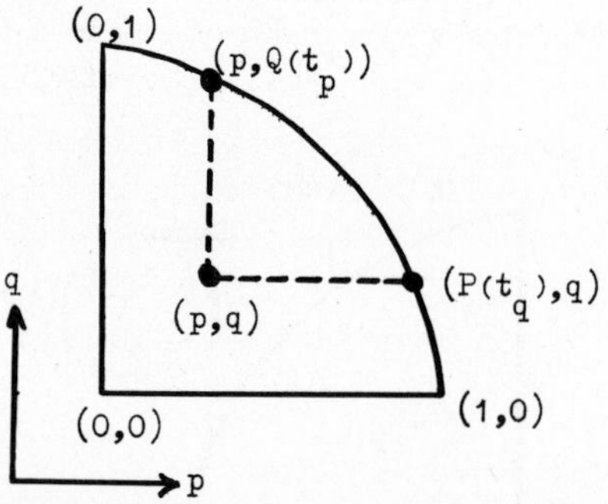

Figure 2.2: Barnhill-Gregory interpolant defined by (2.3).

<u>Nonconforming Linear Element for Curved Triangles</u>

We now give a new curved finite element which has the properties that it matches up with the linear element described in Section 1 (see Figure 1.1). Consider a curved boundary triangle and an adjacent interior straight-sided triangle (Figure 2.3).

We want the curved element to be linearly discretised along its straight sides in such a way that it uses the same parameters as its adjacent straight-sided element. By (1.7) the patch test will then be passed. Moreover, we want to interpolate to all the boundary data. If ——— denotes interpolation all along a curve and ---- denotes a discretisation, then Figure 2.3 becomes Figure 2.4.

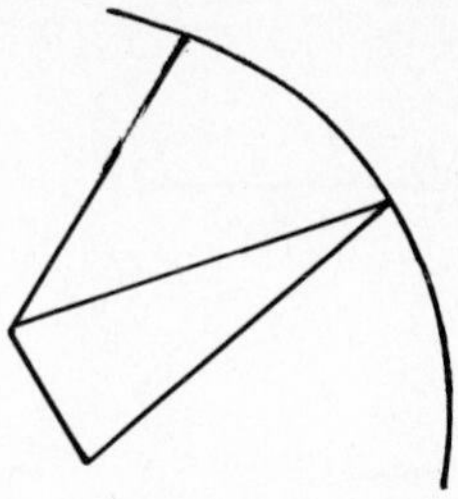

Figure 2.3: Curved boundary triangle and adjacent interior triangle.

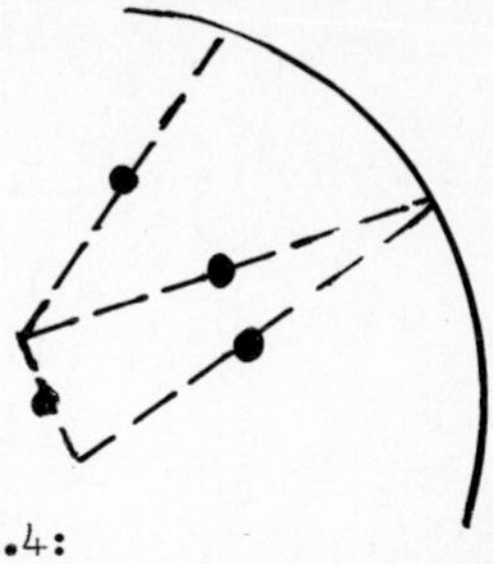

Figure 2.4: Nonconforming linear discretisation of element in Figure 2.3.

Let $L(p,q)$ be the linear interpolant in Section 1. In cardinal form, $L(p,q)$ is given by:

$$L(p,q) = G(\tfrac{1}{2},0)\, Q_1(p,q) + G(0,\tfrac{1}{2})Q_2(p,q) + G(P,Q)Q_3(p,q) \qquad (2.4)$$

where

$$Q_1(p,q) = \frac{1}{D}\{2(2Q-1)P - 4Pq + 2P\}$$

$$Q_2(p,q) = \frac{1}{D}\{-4Qp + 2(2P-1)q + 2Q\}$$

$$Q_3(p,q) = \frac{1}{D}\{2p + 2q - 1\}$$

$$D = 2P + 2Q - 1 .$$

Then an appropriate discretisation of (2.3) along $p = 0$ and $q = 0$ is to replace $G(0,q)$ by $L(0,q)$, etc. This discretisation of (2.3) replaces the values along $p = 0$ and $q = 0$ by the corresponding values of the linear interpolant to G at $(\frac{1}{2},0)$, (P,Q), and $(0,\frac{1}{2})$ (see Figure 2.5).

Figure 2.5: Nonconforming linear discretisation of equation (2.3).

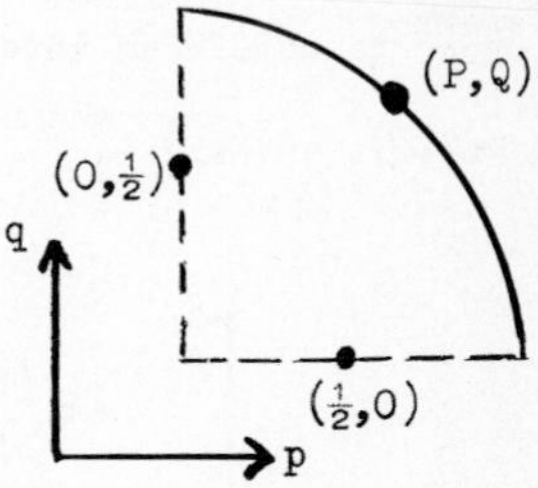

The resulting interpolant is given by equation (2.5).

$$
\begin{aligned}
\tilde{U}(p,q) = {} & qG(p,Q(t_p)) + [1 - Q(t_p)]G(P(t_q),q) \\
& + G(\tfrac{1}{2},0)\{Q(t_p)(\tfrac{-4Pq+2P}{D}) + (1-q)(\tfrac{2(2Q-1)p+2P}{D}) \\
& + [q - Q(t_p)](\tfrac{2P}{D}) - q(\tfrac{-4PQ(t_p)+2P}{D}) - [1 - Q(t_p)](\tfrac{2(2Q-1)P(t_q)+2P}{D})\} \\
& + G(0,\tfrac{1}{2})\{Q(t_p)(\tfrac{2(2P-1)q+2Q}{D}) + (1-q)(\tfrac{-4Qp+2Q}{D}) \\
& + [q - Q(t_p)](\tfrac{2Q}{D}) - q(\tfrac{2(2P-1)Q(t_p)+2Q}{D}) - [1 - Q(t_p)](\tfrac{-4QP(t_q)+2Q}{D})\} \\
& + G(P,Q)\{Q(t_p)(\tfrac{2q-1}{D}) + (1-q)(\tfrac{2p-1}{D}) \\
& + [q - Q(t_p)](-\tfrac{1}{D}) - q(\tfrac{2Q(t_p)-1}{D}) - [1 - Q(t_p)](\tfrac{2P(t_q)-1}{D})\}
\end{aligned}
\tag{2.5}
$$

To reiterate, $\tilde{U}(p,q)$ interpolates to $G(\frac{1}{2},0)$, $G(0,\frac{1}{2})$ and to G all along the curved side (see Figure 2.6)

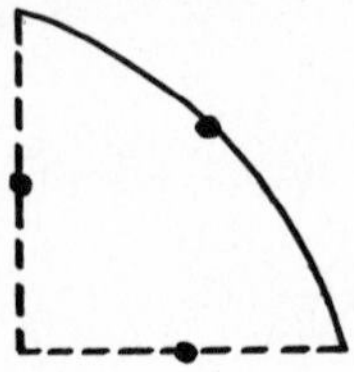

Figure 2.6: Representation of interpolant in equation (2.5).

It is becoming well-known that many standard finite elements can be obtained by the discretisation of an appropriate blending function interpolant. All of these have been conforming elements and the discretisations have been one-dimensional ones along sides of triangles and/or rectangles. The above interpolant thus results from a new kind of discretisation, namely one in which, e.g., $G(0,q)$ depends on values in the whole triangle, not just along $p = 0$.

In order to obtain an interpolant in the physical x-y triangle, one considers $\tilde{U}(p,q)\big|_{p=p(x,y),q=q(x,y)}$ where $p(x,y)$ and $q(x,y)$ are given in (2.2).

<u>Example</u>

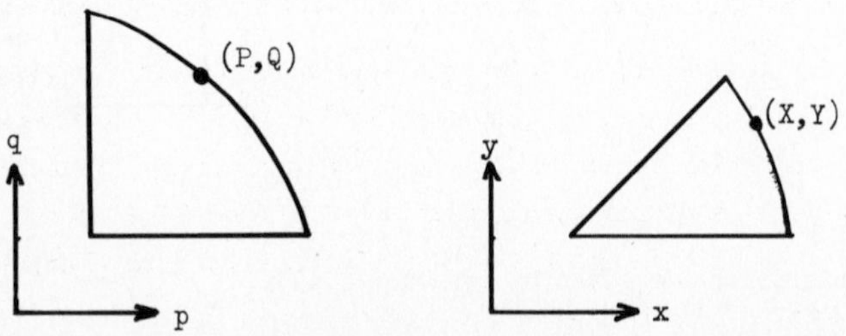

Figure 2.7: Standard and physical triangles, respectively, for the Example.

The physical triangle is a segment of the circle $x^2+y^2 = 1$ with curved side given by

$$\begin{pmatrix} x(t) \\ y(t) \end{pmatrix} = \begin{pmatrix} \cos t \\ \sin t \end{pmatrix} , \qquad 0 \leq t \leq \pi/4 .$$

Let $(X,Y) = (\cos \pi/8, \sin \pi/8) \simeq (0.92, 0.39)$, so that $(P,Q) = (P(\pi/8),Q(\pi/8)) \simeq (0.53, 0.68)$. The interpolant $\tilde{U}(p,q)$ in (2.3) can be calculated in a straightforward manner. Its image in the physical triangle, in cardinal form, is the following:

$$\begin{aligned} V(x,y) = &\sqrt{2}yF(x-y + \sin (\cos-\sin)^{-1}(x-y), \sin(\cos-\sin)^{-1}(x-y)) \\ &+ [1 - \sqrt{2} \sin (\cos-\sin)^{-1}(x-y)]F((\cos-\sin)(\sin^{-1}y) + y,y) + \\ F(\tfrac{1}{2},0) &\left\{ \begin{array}{l} (1-\sqrt{2}y) \frac{1}{D} \{2(2Q-1)(x-y) + 2P\} \\ - [1 - \sqrt{2} \sin(\cos-\sin)^{-1}(x-y)]\frac{1}{D}\{2(2Q-1)(\cos-\sin)\sin^{-1}y + 2P\} \end{array} \right\} + \\ F(\tfrac{1}{2\sqrt{2}}, \tfrac{1}{2\sqrt{2}}) &\left\{ \begin{array}{l} (1-\sqrt{2}y) \frac{1}{D}\{-4Q(x-y) + 2Q\} \\ - [1 - \sqrt{2} \sin(\cos-\sin)^{-1}(x-y)] \frac{1}{D}\{-4Q(\cos-\sin)(\sin^{-1}y) + 2Q\} \end{array} \right\} + \\ F(.92,.39) &\left\{ \begin{array}{l} (1-\sqrt{2}y) \frac{1}{D}\{2(x-y) - 1\} \\ - [1 - \sqrt{2} \sin(\cos-\sin)^{-1}(x-y)]\frac{1}{D}\{2(\cos-\sin)(\sin^{-1}y) - 1\} \end{array} \right\} \end{aligned} \qquad (2.6)$$

where $(\cos t - \sin t)^{-1} = - \int_1^t \frac{dz}{\sin z + \cos z}$.

Curved Morley Triangle

A clamped plate involves a fourth order problem with function and normal derivative given as boundary conditions. A simply supported plate is the same mathematically except that the normal derivative is replaced by a natural boundary condition which need not be built into the approximation. Thus the methods to be given for the clamped plate can be specialised to the simply supported plate.

The normal derivatives to be interpolated are defined on the physical triangle. The Barnhill-Gregory interpolant (equations (5.8)-(5.11) in [2]) interpolates to function values and normal derivatives on the standard curved triangle in the p-q plane (Figure 2.1). The affine transformations (2.1) and (2.2) connect the two curved triangles, but the use of (2.1) and (2.2) presents a problem because normal derivatives are not affine invariant.

The same problem occurs whenever normal derivatives are involved in an interpolant. A general solution is to work on a standard triangle with directional derivatives, the directions to be the images under (2.2) of the directions of the normals in the physical triangle. Basis functions can be tabulated for these directional derivatives on a standard triangle. This procedure will be detailed in Barnhill and Brown [1].

We specialise to the curved Morley triangle. The parameters for the standard Morley triangle with curved sides are depicted in Figure 2.8 (see also Section 1).

Figure 2.8: Curved Morley triangle with normal derivatives.

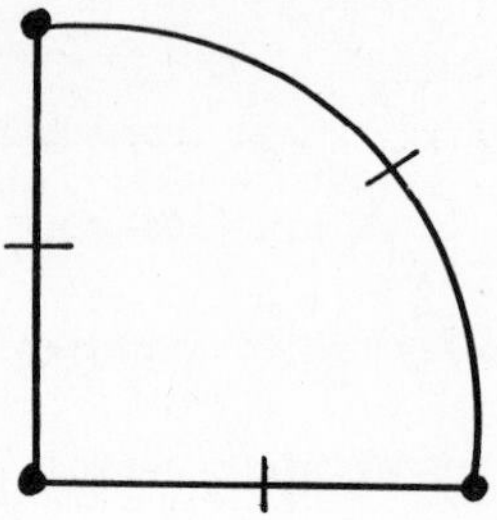

The curved triangle in the physical plane and its affine image, the standard curved triangle, are in Figure 2.9.

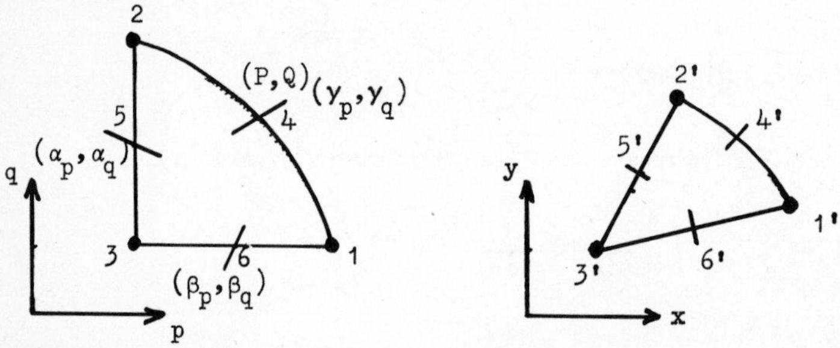

Figure 2.9: Curved Morley triangle with directional derivatives in the p-q plane and with normal derivatives in the physical x-y plane.

As we noted above, the normal derivatives in the physical plane are affinely transformed to directional derivatives in the standard curved triangle. The quadratic function which interpolates to these six parameters is

$$\begin{aligned} M(p,q) &= G(1,0)Q_1(p,q) + G(0,1)Q_2(p,q) + G(0,0)Q_3(p,q) \\ &+ (D_\gamma G)(P,Q)Q_4(p,q) + (D_\alpha G)(0,\tfrac{1}{2})Q_5(p,q) + (D_\beta G)(\tfrac{1}{2},0)Q_6(p,q). \end{aligned} \tag{2.7}$$

The cardinal functions $Q_i(p,q)$ are tabulated in [1]. The finite-dimensional interpolant $M(p,q)$ is substituted into those parts of the Barnhill-Gregory interpolant (equations (5.8) - (5.11) in [2]) that involve function values and derivatives along the straight sides $p = 0$ and $q = 0$ of the standard curved triangle. The derivatives along the curved side are rewritten in terms of the tangential derivative (invariant under affine transformations) and the image $D_\gamma G$ of the normal derivative in the physical triangle. The algebraic details will be given in Barnhill and Brown [1]. The final interpolant for the clamped plate is represented in Figure 2.10.

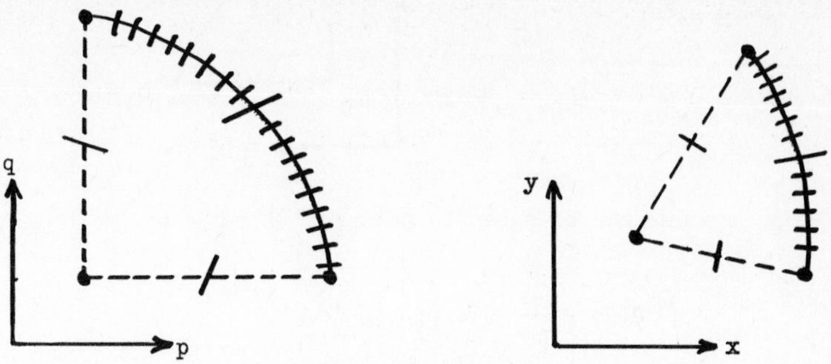

Figure 2.10: Curved Morley element in standard and physical planes, respectively.

Interpolation to function value and normal derivative all around the curved side is guaranteed by Theorem 2.2, part (i), of Barnhill and Gregory [2].

The analogous interpolant for the simply supported plate is represented in Figure 2.11 (see [1]).

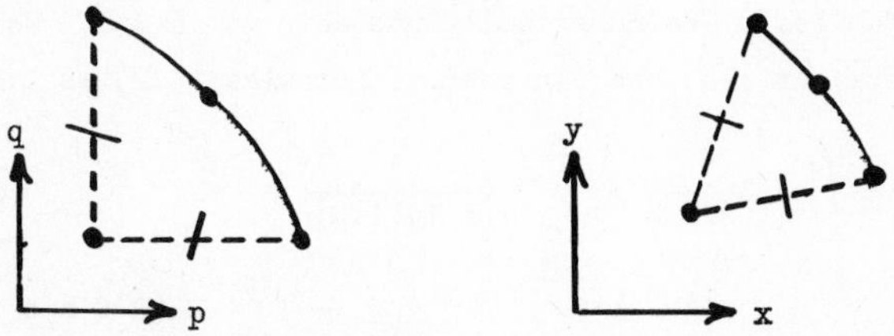

Figure 2.11: Curved Morley element for simply supported plate.

3. Numerical Integration

As well as the general non-conforming curved elements constructed above, various types of conforming elements for second order problems have also been produced, e.g. McLeod and Mitchell [7]. As with all Ritz-Galerkin approximations, it is necessary to evaluate certain integrals over each element, and in particular over each curved element. In this section we consider the problem of computing the required integrals numerically for a particular choice of the curved side.

The simplest non-trivial choice of curved side is a hyperbola. We assume that the hyperbola has equation

$$1 - p - q + bpq = 0 \tag{3.1}$$

where $b = (P+Q-1)/PQ$ and (P,Q) is a point on the hyperbola (see Figure 3.1).

Figure 3.1: Triangle with hyperbolic curved side.

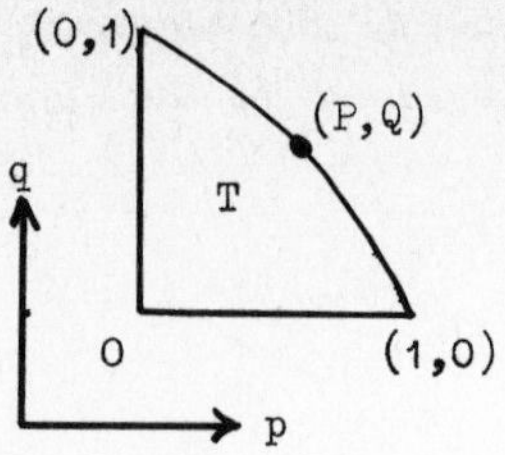

The following formulae can be shown to hold for $0 < b < 1$:

$$\iint_T dpdq = \frac{1}{b^2}\{b + (1-b)\ln(1-b)\}$$

$$\iint_T pdpdq = \iint_T qdpdq = \frac{1}{b^3}\{b - \tfrac{1}{2}b^2 + (1-b)\ln(1-b)\} \qquad (3.2)$$

As we increase the number of elements, we would expect that the hyperbola would become more and more like the straight line $1 - p - q = 0$ i.e. $b \to 0$. It is obvious that the formulae (3.2) are numerically unstable as $b \to 0$. Using Maclaurin's expansion for $\ln(1-b)$, we can rewrite formulae (3.2) as

$$\iint_T dpdq = \tfrac{1}{2} + \sum_{i=1}^{\infty} \frac{b^i}{(i+1)(i+2)}$$

$$\iint_T pdpdq = \iint_T qdpdq = \frac{1}{6} + \sum_{i=1}^{\infty} \frac{b^i}{(i+2)(i+3)} \qquad (0 \leqslant b \leqslant 1) \qquad (3.3)$$

There will be a crossover value of b, say b^*, when:

(i) $b > b^*$ => formulae (3.2) satisfactory as regards round off and stability.

(ii) $b \leqslant b^*$ => formulae (3.3) should be used.

In using (3.3) we would, in practice, use as few terms as possible - until, for example, the integrals are correctly evaluated to machine accuracy (or even less precisely than this, to some chosen number of decimal places).

Using the above formulae it is easy to construct an integration formula with linear precision. We choose w^*, p^*, and q^* so that

$$w^* = \iint_T dpdq$$

$$w^*p^* = \iint_T pdpdq \qquad (3.4)$$

$$w^*q^* = \iint_T qdpdq$$

and we have the simple integration rule:

$$\iint_T f(p,q)dpdq \simeq w^*f(p^*,q^*) \qquad (3.5)$$

with (i) positive weight w^*, (ii) interior point (p^*,q^*) - both desirable features.

It is not as easy to produce rules with higher order precision. For such cases the resulting system of non-linear equations may be difficult (or impossible) to solve analytically. This is still an open problem. There is, however, a dual technique, viz. integration by interpolation, which can be used. This approach will be detailed in Barnhill and Brown [1].

Acknowledgements. The research of R E Barnhill was supported by The Science Research Council with Grant B/RG/61876 to The University of Dundee and by The National Science Foundation with Grant DCR74-13017 to The University of Utah. This author wishes to give particular thanks to Professor A R Mitchell for setting up his year at The University of Dundee. The research of J H Brown was carried out while in receipt of a Science Research Council Research Studentship.

REFERENCES

1. R E Barnhill and J H Brown, Curved Nonconforming Elements for Plate Problems, Numerical Analysis Report No.8, University of Dundee Mathematics Dept., 1975.

2. R E Barnhill and J A Gregory, Polynomial Interpolation to Boundary Data on Triangles, Math. Comp. 29, 726-735, 1975.

3. J H Brown, Conforming and Nonconforming Finite Element Methods for Curved Regions Ph.D. Thesis, University of Dundee (to appear).

4. P G Ciarlet, Conforming and Nonconforming Finite Element Methods for Solving the Plate Problem, Dundee Conference Proceedings, G A Watson (ed.), Springer-Verlag 1973.

5. P G Ciarlet, Numerical Analysis of the Finite Element Method, Séminaire de Mathématiques Supérieures, Université de Montréal, 1975.

6. B M Irons and A Razzaque, Experience with the Patch Test for Convergence of Finite Element Methods, The Mathematical Foundations of the Finite Element Method with Applications to Partial Differential Equations, A K Aziz (ed.), Academic Press, 1972.

7. R J Y McLeod and A R Mitchell, The Construction of Basis Functions for Curved Elements in the Finite Element Method, J.I.M.A. 10, 382-393, 1972.

8. L S D Morley, The Triangular Equilibrium Element in the Solution of Plate Bending Problems, Aero. Quart. 149-169, 1968.

9. L S D Morley, A Triangular Equilibrium Element with Linearly Varying Bending Moments for Plate Bending Problems, J Royal Aero. Soc., 71, 715-719, 1967.

10. G Strang, Variational Crimes in the Finite Element Method (see Ref.6).

11. G Strang and G J Fix, An Analysis of the Finite Element Method, Prentice-Hall, 1973.

12. B Fraeijs de Veubeke, Variational Principles and the Patch Test, Int. Jour. Num. Meth. Eng., 8, 783-801, 1974.

THE APPROXIMATE SOLUTION OF LINEAR AND NONLINEAR FIRST-KIND INTEGRAL EQUATIONS OF VOLTERRA TYPE

Hermann BRUNNER

I. Introduction

Let T denote the Volterra integral operator defined by

$$(1.1)\qquad (Tf)(x) = \int_a^x K(x,t,f(t))\,dt\ , \quad x \in I = [a,b] \quad (I \text{ compact})\ ,$$

with $T: C(I) \longrightarrow C_a(I)$, where $C_a^r(I) = \{f \in C^r(I) : f(a) = 0\}$, $C^0 = C$. More precise conditions on T will be given in Section 2; we note here that the kernel K of T may possess a weak singularity along $x = t$, and the term Volterra operator is meant to include operators with weakly singular kernels of Abel type.

For a given function $g \in C_a^r(I)$ $(r \geq 0)$ the integral equation

$$(1.2)\qquad (Ty)(x) = g(x)\ , \quad x \in I\ ,$$

is to be solved in the space C(I). We shall always assume that (1.2) possesses a unique exact solution $y \in C(I)$; for a discussion of existence and uniqueness questions for equations of the form (1.2) compare, for example, [17] (see also [14] for results on nonlinear generalized Abel equations).

An explicit expression for the exact solution of (1.2) exists only for a very few special cases (we mention the classical Abel equation, linear equations with $K(x,t,u) = k(x-t)u$ (under certain conditions on k), and the trivial case $K(x,t,u) = p(x)q(t)u$ (with sufficiently smooth $p,q \neq 0$ on I)). In general, therefore, one has to rely on numerical methods for generating an approximate solution for (1.2);

this also holds in the classical Abel case if (as is often the situation in equations arising from physical problems) $g(x)$ is not known explicity but is only given on some finite subset of I.

In the present paper we shall be interested in certain direct methods for numerically solving (1.2) (i.e. methods which do not make use of an inversion formula for the exact solution and which are not based on the transformation of (1.2) into an equation of the second kind). These methods may be regarded as projection methods (in general nonlinear), and they yield an approximation for $y(x)$ on I rather than on some discrete subset of I. The motivation for considering such methods comes from the fact that direct methods based on numerical quadrature (or, more generally, on discretization) have a number of severe drawbacks. Recall that methods of this type (as well as methods based on Runge-Kutta techniques and related methods; see [10], [11], [21], for example) use the following basic idea: I is replaced by a subset $Z_N = \{x_{N,k} = a + kh_N : k = 0,1,\dots,N;\ x_{N,N} = b\}$, and the operator T is then replaced by a sequence of operators $\{T_n^{(N)} : n = r,\dots,N;\ r \geq 1\}$ (for example, by selecting a family of quadrature functionals $\{Q_n^{(N)}\}$, with

$$Q_n^{(N)} : \int_a^{x_{N,n}} f(t)dt \longrightarrow \sum_{j=0}^{n} w_{n,j}^{(N)} f(x_{N,j}) \ , \ n = r,\dots,N \ ,$$

and by setting

$$(T_n^{(N)}u)(x) = \sum_{j=0}^{n} w_{n,j}^{(N)} K(x, x_{N,j}, u(x_{N,j})) \quad);$$

one then solves the algebraic system

$$(T_n^{(N)}u)(x) = g(x) \quad \text{for} \quad x \in Z_N^{(r)} = \{x_n \in Z_N : n \geq r\} \ .$$

Among the drawbacks mentioned above are: (i) divergence of higher-order (multistep) methods based on numerical quadrature, (ii) generation of additional starting values, (iii) a change of stepsize during the computational process is difficult, (iv) no uniform treatment of singular and nonsingular equations is possible, (v) non-smooth K and/or g dictate excessively small stepsizes (even if g, for example, is non-smooth only on a small part of I). For details we refer to [19], [18], [21], [10], [11], [8], [12] .

The approach taken in this paper is based on the following idea. The physical problem from which a given integral equation of the form (1.2) results often yields some information about the function space in which the (unknown) exact solution $y(x)$ lies. Hence let $V \subset C(I)$ be chosen accordingly, with dim V finite (precise criteria for the choice of V will be given in Section 2). Instead of replacing the operator T (as in discretization methods or in kernel approximation methods (see [9], [2]), T is retained and one considers the problem of minimizing (in a prescribed sense) the expression $T\Phi - g$ $(\Phi \in V)$, where the solution is sought in the subspace $W = TV \subset C_a(I)$. If $g \in W$ then, of course, the approximate solution Φ satisfies $\Phi = y$.

The mathematical motivation for the approach sketched above is to be found in the fact that so-called positive Volterra operators (defined in Section 2) leave crucial structures of approximating families (essentially) invariant. This will be made precise in the following section.

2. Approximation properties of positive Volterra operators

It will be assumed in the following that the Volterra operator T defined by (1.1) possesses the properties listed below.

(i) $K \in C(S)$, $S = S_1 \times \mathbb{R}$, $S_1 = \{(x,t) : a \le t \le x \le b\}$;

(ii) $K_1 = K_1(x,t,u) = \frac{\partial}{\partial u} K(x,t,u) \in C(S)$;

(iii) $|K_1(x,t,u)| \geq \rho > 0$ for $(x,t,u) \in S$.

If the kernel of T is of the form $K(x,t,u) = G(x,t,u)(x-t)^{-\alpha}$, $\alpha \in (0,1)$, then the function G is assumed to take the rôle of K in (i) - (iii).

<u>DEFINITION:</u> If a (nonsingular or weakly singular) Volterra operator of the form (1.1) has the properties (i) - (iii) it is called a positive Volterra operator.

Volterra operators with kernels of this type (which occur in most of the integral equations arising from physical problems) leave, as has already been indicated, certain crucial structures of approximating families (like the (local and global) Haar property and asymptotic convexity (which will not be needed in the present context)) essentially invariant. The precise results are contained in Theorems 1a and 1b below. The following definition is a modification of a terminology introduced in [6].

<u>DEFINITION:</u> Let $Q \subset I$ consist of a finite number of points, and let $V \subset C(I)$ have (finite) dimension m. The subspace $V = V_Q$ is called a Haar subspace (on I) with (discrete) weak null space Q if any nontrivial $\Phi \in V$ has at most (m-1) zeros on $I-Q$ and does not change sign on Q.

In practical applications one will usually be concerned with the case $Q \subseteq \{a,b\}$; $Q = \emptyset$ corresponds to the case of a classical Haar subspace of $C(I)$.

<u>THEOREM 1a:</u> Let $V = V_Q \subset C(I)$ be a Haar subspace (on I) of dimension m and with (discrete) weak null space $Q \subset I$, and let T be a positive

Volterra operator. Then the nonlinear family $W = TV \subset C_a(I)$ satisfies the local Haar condition on $(a,b]$; that is, each nontrivial element from

$$H = H(\alpha) = \text{span}\ \{\frac{\partial}{\partial\alpha_j}\Psi(\alpha,x)\ :\ j = 1,\dots,m;\ \Psi \in W\}$$

(with $\dim H(\alpha) = \dim V = m$ for all $\alpha \in R^m$) vanishes at most $(m-1)$ times on $(a,b]$.

<u>Proof:</u> If $\Phi = \Phi(\alpha,x) = \sum_{\nu=1}^{m} \alpha_\nu \phi_\nu(x) \in V (\Phi \not\equiv 0)$ then, by hypotheses (i) - (iii),

$$h_j(\alpha,x) = \frac{\partial}{\partial\alpha_j}(T\Phi)(\alpha,x) = \int_a^x K_1(x,t,\Phi(\alpha,t))\cdot\phi_j(t)dt,\ j = 1,\dots,m\ ,$$

with $h_j(\alpha,a) = 0, \alpha \in R^m$. Since K_1 is nonzero on S and by assumption on V and Q, an elementary counting argument yields the desired result: in

$$h(\beta\ ;\alpha,x) = \sum_{j=1}^{m} \beta_j h_j(\alpha,x) = \int_a^x K_1(x,t,\Phi(\alpha,t))\cdot\Phi(\beta,t)dt\ ,$$

only a sign change of $\Phi(\beta,t)$ can generate a zero on $(a,b]$.

<u>THEOREM 1b:</u> Let V and T be as in Theorem 1a. Then the set $W = TV$ satisfies the global Haar condition on $(a,b]$; that is, for any $\alpha,\beta \in R^m$ and with $\Psi = T\Phi$, $\Phi \in V$, either $\Psi(\alpha,x) - \Psi(\beta,x) \equiv 0$, or $\Psi(\alpha,x) - \Psi(\beta,x)$ vanishes less than m times on $(a,b]$.

<u>Proof:</u> By assumption (i) and by linearity of V we have

$$\Psi(\beta,x) - \Psi(\alpha,x) = (T\Phi)(\beta,x) - (T\Phi)(\alpha,x) =$$

$$= \int_a^x K_1(x,t,\Phi(\alpha + \theta(\beta-\alpha),t))\cdot\Phi(\beta-\alpha,t)dt \quad (\theta \in (0,1)\ )\ .$$

The result now follows from the positivity of T.

It also follows from the above proof that if (1.2) (with positive operator T) possesses a solution $y \in C(I)$ it must be unique.

If the phrase "positive Volterra operator T" is replaced by "positive Fredholm operator F", with $(Ff)(x) = \int_a^b K(x,t,f(t))dt$ and with K satisfying (i) - (iii) (where S_1 is replaced by $I \times I$), then the above results are no longer true in general. To see this, let $K(x,t,u) = (1 + q(x)t)u$, with $q \in C(I)$, $q(x)t \geq c > -1$ on $I \times I$. Then, for $V = \pi_1$, $\Psi(\alpha,x) = (3\alpha_1 + 2\alpha_2)\cdot q(x)/6 + (2\alpha_1 + \alpha_2)/2$, with $\Phi(\alpha,x) = \alpha_1 + \alpha_2 x$. If $2\alpha_1 + \alpha_2 = 0$ $(\alpha_1 \cdot \alpha_2 \neq 0)$, then $\Psi(\alpha,x) \not\equiv 0$. But Ψ clearly vanishes at zeros of $q(x)$. Hence, for a function like $q(x) = \sin(n\pi x)/2$, $n > 1$, the space $W = FV$ is not a Haar subspace of dimension $m = 2$ on $I = [0,1]$. We note, however, that for linear Fredholm operators we have the classical result that F transforms a (classical) Haar subspace of dimension m into a Haar subspace of the same dimension if its kernel satisfies

$$k\begin{pmatrix} x_1 & \cdots & x_m \\ t_1 & \cdots & t_m \end{pmatrix} > 0 \tag{2.1}$$

for any points $a \leq x_1 < \ldots < x_m \leq b$, $a \leq t_1 < \ldots < t_m \leq b$ (see [7] , pp. 160-161).

<u>THEOREM 2:</u> Let T be a positive Volterra operator, and suppose that T is subject to the additional hypothesis

$$\lim_{|u| \to \infty} |K(x,t,u)| = \infty \quad \text{for all} \quad (x,t) \in S_1 . \tag{2.2}$$

If $g \in C_a(I)$, and if $V = V_Q \subset C(I)$ is a Haar subspace with (discrete) weak null space $Q \subset I$, then there exists one (and hence only one) best Chebyshev approximation to g on I from the (nonlinear) set $W = TV \subset C_a(I)$.

The proof of this result will appear elesewhere; we note that uniqueness follows (by a slight modification of the standard arguments, since we approximate in $W \subset C_a(I)$) from Theorem 1a and Theorem 1b.

If T is linear, i.e. $K(x,t,u) = k(x,t)u$ with $k \in C(S_1)$ or, if $k(x,t) = \kappa(x,t)(x-t)^{-\lambda}$, $\lambda \in (0,1)$, with $\kappa \in C(S_1)$, then positivity of $K_1(x,t,u) = k(x,t)$ on S_1 implies that (2.2) holds. We also observe that Theorem 2 will be valid for (linear) Fredholm operators F (where now $g \in C(I)$) and the corresponding equation (1.2) if the kernel of F satisfies, as indicated above, condition (2.1).

3. Linear operators: error bounds and convergence

In this section we indicate a typical convergence result for an equation (1.2) corresponding to a linear (weakly singular) positive Volterra operator; analogous results for nonsingular operators (with weaker condition on K and g) may be obtained in a similar fashion. The weakly singular case has been selected also to show that often knowledge about certain properties of g in (1.2) can be used when choosing the (Haar) subspace V in which the approximate solution is sought.

Let $\Psi^* = \Psi(\alpha^*,x) \in W = TV$ (with $\Psi(\alpha^*,x) = (T\Phi)(\alpha^*,x)$) denote some approximation to g on I (obtained, for example, by Chebyshev or L_1-approximation, or by interpolation in W), and set $e^* = e(\alpha^*,x) = y(x) - \Phi(\alpha^*,x)$, $\eta^* = \eta(\alpha^*,x) = (Te)(\alpha^*,x)$. We then have (with $\|f\| = \max\{|f(x)| : x \in I\}$, $f \in C(I)$)

<u>THEOREM 3:</u> Suppose that T is a positive Volterra operator given by $K(x,t,u) = \kappa(x,t)(x-t)^{-\lambda}u$, $\lambda \in (0,1)$, with $\kappa^{(1)}(x,t) = \frac{\partial}{\partial x}\kappa(x,t) \in C(S_1)$, $|\kappa^{(1)}(x,t)| \leq M_1$ on S_1, and let $g \in C_a^1(I)$. Then:

(a) The (unique) exact solution $y \in C(I)$ of the corresponding (1.2) is in $C_a(I)$.

(b) If $V \subset C_a(I)$, the error e^* induced by the approximation $\Phi \in V$ satisfies

$$\|e^*\| \leq \frac{\sin(\lambda\pi)}{\lambda\pi}(b-a)^{\lambda} \cdot \rho^{-1} \exp((1-\lambda)M_1 \cdot \rho^{-1}(b-a)) \cdot \|(\eta^*)'\|, \tag{3.1}$$

where $' = d/dx$.

Proof: It is not difficult to show that, under the above hypotheses, e^* is the solution of the integral equation [17]

$$e(\alpha^*,x) = \frac{\sin(\lambda\pi)}{\kappa(x,x)\pi}\int_a^x \eta'(\alpha^*,t)(x-t)^{\lambda-1}dt \; +$$

$$+ \frac{\sin(\lambda\pi)}{\kappa(x,x)\pi}\int_a^x \int_0^1 \left(\frac{r}{1-r}\right)^{1-\lambda} \cdot \kappa'(t+r(x-t),t)dr \cdot e(\alpha^*,t)dt, \; x \in I .$$

The result follows by taking absolute values, by observing that $|\kappa(x,t)| \geq \rho > 0$ on S_1 and that $\int_0^1 \left(\frac{r}{1-r}\right)^{1-\lambda} dr = B(2-\lambda,\lambda) = \Gamma(\lambda)\Gamma(2-\lambda) = (1-\lambda)\pi/\sin(\lambda\pi)$, and by then applying Gronwall's inequality.

Theorem 3 may be used to establish convergence: if $V = V^{(m)} \subset C_a(I)$, with $m = \dim V \to \infty$, are selected such that $\eta(\alpha^*,x) \to 0$ uniformly on I, then $\Phi(\alpha^*,x) \to y(x)$ uniformly on I.

The above result also suggests that the choice $V = V^{(m)} = \pi_{m-1}$, $m \to \infty$, will not yield convergence in general if Φ^* is determined by interpolation (collocation) by $\Psi^* = T\Phi^*$ to g on given finite subsets $Z_m \subset (a,b]$. On the other hand, if y is approximated by piecewise polynomials of class $C(I)$ on a chosen partition of I (by recursive interpolation in the transformed space W) then uniform convergence (which implies numerical stability; a situation resembling the one in the theory of finite-elements for partial differential equations

([20], p. 17)) occurs (compare [4] for details).

4. Interpolation and Chebyshev approximation

Suppose that T is again a linear positive Volterra operator, and let $V = V_Q \subset C(I)$ denote a Haar subspace of dimension m and with (discrete) weak null space $Q \subset I$. If $Z_m = \{z_k : a < z_1 < \ldots < z_m \leq b\}$ is given, then it follows from Theorem 1a (or 1b) that the interpolation problem in $W = TV$ (collocation),

(4.1) $\qquad \Psi(\alpha,x) = (T\Phi)(\alpha,x) = g(x), \quad x \in Z_m \quad (g \in C_a(I))$

has a unique solution. It is natural, therefore, to ask for the "optimal" set $Z_m^* \subset (a,b]$, i.e. the set Z_m^* whose corresponding solution Φ^* of (4.1) satisfies

(4.2) $\qquad \|y - \Phi^*\| = \inf \{\|y - \Phi\| : \Phi \in V\}$.

Unfortunately, it turns out that in general such a set Z_m^* does not exist. As a simple example, consider the interval $I = [-1,1]$, with $V = \pi_{m-1}$, $K(x,t,u) = u$, and let $y \in \pi_m$. The error function $e = e(\alpha,x)$ corresponding to the best uniform approximation Φ to y from V has the form $e(\alpha,x) = c \cdot T_m(x)$ (where T_m denotes the first-kind Chebyshev polynomial of degree m). For the existence of an optimal set Z_m^* described above (assuming that $\Phi \notin \pi_{m-2}$) it is necessary that $\int_{-1}^{x} T_m(t)dt$, $x \in I$, possesses m distinct zeros in $(-1,1]$. It is easily verified that this will hold if, and only if, m is odd (the largest zero is then $z_m = 1$); if m is even, only $(m-1)$ zeros lie in $(-1,1]$.

On the other hand, let again $K(x,t,u) = u$, $I = [-1,1]$. Choose $y(x) = x^m - \sigma \cdot x^{m-1}$ $(m \geq 2;\ \sigma \geq 0)$, and let $V = \pi_{m-2}$. Here,
$e = e(\alpha,x) = 2^{1-m}(1 + \sigma/m)^m \cdot T_m\left(\frac{x-\sigma/m}{1+\sigma/m}\right)$, $x \in I$ $\quad (0 \leq \sigma \leq \tan^2(\pi/2m)$;

e is the well-known Zolotareff polynomial of degree m). It can be shown without difficulty that, for $0 \leqq \sigma \leqq \tan^2(\pi/4m)$, the function $\int_{-1}^{x} T_m(\frac{t-\sigma/m}{1+\sigma/m})dt$ possesses $(m-1)$ distinct zeros in $(-1,1]$ for any $m \geqq 2$. Hence an optimal set Z^*_{m-1} for (4.1) exists (since dim $V = m-1$) for sufficiently small values of σ.

At present it is not known (even for the simple case $K(x,t,u)=u$) under what condition on g (for a given kernel K and a prescribed space $V = V_Q \subset C(I)$) one can guarantee the existence of an optimal set Z^*_m of interpolation points in (4.1)(note that in general the best Chebyshev approximation to y from V_Q will not be unique). Equivalently: if V and K are given, when does the integral of a best (Chebyshev) error curve, $\int_a^x K(x,t)\cdot e(\alpha\ ,t)dt$, $x \in I$, possess m distinct zeros in $(a,b]$?

5. Some numerical aspects and open problems

It has become clear from the previous discussion that, in general, the space $W = TV$ (in which the approximation problem for the given $g \in C_a(I)$ in (1.2) is to be solved) will not be known explicitly. (For many types of kernels, however, an appropriate choice of the space V will make it possible to get W explicitly; we mention, as an example, the case $K(x,t,u) = (x-t)^{-\lambda}\cdot u$ (Abel kernel) with $V = \text{span}\,\{x^{p_1}, \ldots, x^{p_m}\}$, $0 \leqq p_1 < \ldots < p_m$.) On the other hand, the function $\Psi(\alpha,x) = \int_a^x K(x,t,\Phi(\alpha,t))dt$ (or, if T is linear, the basis functions $\psi_\nu(x) = \int_a^x k(x,t)\phi_\nu(t)dt$, of W, where $V = \text{span}\,\{\phi_1, \ldots, \phi_m\}$) can be computed, for any $x \in I$ and any $\alpha \in R^m$, to a prescribed accuracy by selecting one of the many available modern numerical integration subroutines. If the integrand is smooth then a method based on Gaussian quadrature will be a prime candidate. (We recall that weakly singular integrals with rational λ may be transformed into integrals with regular kernels; compare [3], p. 74.) This situation which usually requires the evaluation of a large number of integrals (especially in the case of Chebyshev approximation on I, when using one of the algorithms of Remez: the search for one or several extrema needed for the exchange step will dictate the knowledge of $\Psi(\alpha,x)$ on some mesh of I) is similar to the one encountered when solving Fredholm integral equations of the second kind by kernel approximation methods with subsequent numerical quadrature (see [2]).

If T is nonlinear, and if the approximation problem for g in $W = TV$ is solved by the Remenz algorithm or by one the related methods then, in contrast to most nonlinear Chebyshev approximation problems, the dimension m of the space $H = H(\alpha)$ (compare Theorem 1a) is independent of α .

We conclude with two open problems (compare also section 4).

I. Let V be a given Haar subspace with (discrete) weak null space $Q \subset I$. Consider those positive Volterra operators T and those functions $g \in C_a(I)$ for which (1.2) possesses a unique solution $y \in C(I)$. How can those T be characterized for which $\Phi^* \in V$ corresponding to the solution Ψ^* of $\|g-\Psi^*\| = \inf \{\|g-\Psi\| : \Psi \in W\}$ satisfies $\|y-\Phi^*\| = \inf \{\|y-\Phi\| : \Phi \in V\}$?

II. Is the conjecture true that, under the assumptions of Theorem 2, the set $W = TV$ is unisolvent?

REFERENCES

1. P.M. Anselone, Collectively Compact Operator Approximation Theory and Application to Integral Equations, Prentice-Hall, Englewood Cliffs (N.J.), 1971.

2. P.M. Anselone and J.W. Lee, Double approximation schemes for integral equations, to appear in: Proc. Confer. Approximation Theory (Math. Research Inst. Oberwolfach (Germany), May 25-30, 1975), Birkhäuser-Verlag, Basel.

3. H. Brunner, On the approximate solution of first-kind integral equations of Volterra type, Computing (Arch. Elektron. Rechnen), 13 (1974), 67-79.

4. H. Brunner, Global solution of the generalized Abel integral equation by implicit interpolation, Math. Comp., 28 (1974), 61-67.

5. C.B. Dunham, Chebyshev approximation with a null point, Z. Angew. Math. Mech., 52 (1972), 239.

6. C.B. Dunham, Families satisfying the Haar condition, J. Approx. Theory, 12 (1974), 291-298.

7. F.R. Gantmacher und M.G. Krein, Oszillationsmatrizen, Oszillationskerne und kleine Schwingungen mechanischer Systeme, Akademie-Verlag, Berlin, 1960.

8. C.J. Gladwin, Numerical Solution of Volterra Integral Equations of the First Kind, Ph.D. Thesis, Dalhousie University, Halifax, N.S., 1975.

9. J. Hertling, Numerical treatment of singular integral equations by interpolation methods, Numer. Math., 18 (1971/72), 101-112.

10. F. de Hoog and R. Weiss, On the solution of Volterra integral equations of the first kind, Numer. Math., 21 (1973), 22-32.

11. F. de Hoog and R. Weiss, High order methods for Volterra integral equations of the first kind, SIAM J. Numer. Anal., 10 (1973), 647-664.

12. P.A. Holyhead, S. McKee and P.J. Taylor, Multistep methods for solving linear Volterra integral equations of the first kind, to appear in: SIAM J. Numer. Anal.

13. Y. Ikebe, The Galerkin method for numerical solution of Fredholm integral equations of the second kind, SIAM Review, 14 (1972), 465-491.

14. J. Janikowski, Équation intégrale non linéaire d'Abel, Bull. Soc. Sci. Lettres Łódź, 13 (1962), no. 11.

15. E.H. Kaufman and G.G. Belford, Transformation of families of approximating functions, J. Approx. Theory, 4 (1971), 363-371.

16. E.L. Kosarev, The numerical solution of Abel's integral equation, Zh. vycisl. Mat. mat. Fiz., 13 (1973), 1591-1596 (= U.S.S.R. Comput. Math. and Math. Phys., 13 (1973), 271-277).

17. G. Kowalewski, Integralgleichungen, de Gruyter, Berlin, 1930.

18. P. Linz, Numerical methods for Volterra integral equations of the first kind, Comput. J., 12 (1969), 393-397.

19. B. Noble, The numerical solution of nonlinear integral equations and related topics, in: P.M. Anselone (Ed.), Nonlinear Integral Equations, University of Wisconsin Press, Madison, 1964: 215-318.

20. G. Strang and G.J. Fix, An Analysis of the Finite Element Method, Prentice-Hall, Englewood Cliffs (N.J.), 1973.

21. R. Weiss, Product integration for the generalized Abel equation, Math. Comp., 26 (1972), 177-190.

A CLASS OF IMPLICIT METHODS FOR ORDINARY DIFFERENTIAL EQUATIONS

J.C. Butcher

1. Introduction

Since implicit methods, typically, have better stability properties than explicit methods, it is of some interest to make a detailed study of one particular class of implicit methods. Specifically, this paper will deal with the class of two-stage implicit Runge-Kutta methods. For the differential equation $y'(x) = f(x,y(x))$ the solution computed at $x_N = x_{N-1} + h$, for h a constant step size, is y_N where

(1)
$$\begin{aligned} k_1 &= hf(x_{N-1}+hc_1,\ y_{N-1}+a_{11}k_1+a_{12}k_2) \\ k_2 &= hf(x_{N-1}+hc_2,\ y_{N-1}+a_{21}k_1+a_{22}k_2) \\ y_N &= y_{N-1}+b_1k_1+b_2k_2 \end{aligned}$$

and $c_1, c_2, a_{11}, a_{12}, a_{21}, a_{22}, b_1, b_2$ are numerical constants. We will always assume that f satisfies a Lipschitz condition so that for sufficiently small $|h|$, y_N is uniquely determined as a function of y_{N-1}. This method will be characterized by an array as follows

(2)
$$\begin{array}{c|cc} c_1 & a_{11} & a_{12} \\ c_2 & a_{21} & a_{22} \\ \hline & b_1 & b_2 \end{array}$$

and it will always be assumed that $c_1 = a_{11} + a_{12}$ and $c_2 = a_{21} + a_{22}$.

The method (2) is known [1] to be of order 4 if and only if its array is

$$\begin{array}{c|cc} \frac{1}{2}-\frac{\sqrt{3}}{6} & \frac{1}{4} & \frac{1}{4}-\frac{\sqrt{3}}{6} \\ \frac{1}{2}+\frac{\sqrt{3}}{6} & \frac{1}{4}+\frac{\sqrt{3}}{6} & \frac{1}{4} \\ \hline & \frac{1}{2} & \frac{1}{2} \end{array}$$

(or the equivalent method where the rôles of k_1 and k_2 are interchanged). In this case, the method is A-stable.

To obtain the class of methods with which this paper is concerned, we replace the condition that the order is four by the more general condition that the "effective order" [2] is four.

2. Definition of Effective Order

For a method m, a step size h and a given function f(that is, a given differential equation) we shall write $\phi(m,h,f)$ for the function that maps an initial value to the result computed after a single step starting from that initial value. Thus if m is given by (2), then y_N given by (1) can be written as $y_N = \phi(m,h,f)(y_{N-1})$.

The method m has associated with it a method m^{-1} given by

$$\begin{array}{c|cc} c_1 - b_1 - b_2 & a_{11} - b_1 & a_{12} - b_2 \\ c_2 - b_1 - b_2 & a_{21} - b_1 & a_{22} - b_2 \\ \hline & -b_1 & -b_2 \end{array}$$

such that

$$\phi(m^{-1},h,f) \circ \phi(m,h,f) = \phi_o(h,f)$$

where ϕ_o is such that for $|h|$ sufficiently small, $\phi_o(h,f)$ maps y_{N-1} to itself. In this sense, m^{-1} can be regarded as the inverse to m.

Furthermore, for any two Runge-Kutta methods m and m_1, there exist a method which will be denoted by mm_1 such that, for sufficiently small $|h|$,

$$\phi(mm_1, h, f) = \phi(m_1,h,f) \circ \phi(m,h,f)$$

We can now state

Definition A method m is of Effective Order n if there exists a method m_1 such that $m_1^{-1}\, m\, m_1$ is of order n.

To make practical use of this definition, the method m has to be implemented in a special way. That is, before the computations with m begin, a preliminary step with m_1^{-1} is performed. This results in a perturbation being applied to the initial value that could, in principle, be removed by an application of the method m_1. This perturbed, or what can be called Butchered, initial value is used as a basis for computations with method m over the required number of steps and then, finally, the perturbation is eliminated by the application of a single step with method m_1.

The result computed in the way described here has error behaviour as $h \to 0$ just as if an nth order Runge-Kutta method were used in the normal way throughout the computation. A formal proof, in a more general setting, is given in [3].

3. Algebraic interpretation of Effective Order

To express the conditions of effective order algebraically, we will make use of results in [4] . In that paper, a certain group G was introduced such that to each Runge-Kutta method there corresponds a unique element of G which, in turn, characterises the method. Furthermore, if $\alpha, \beta \in G$ correspond to methods a,b then $\alpha\beta$ corresponds to ab and α^{-1} corresponds to a^{-1}. One particular element of G, denoted by p, corresponds to the "Runge-Kutta method" (with a continuum of stages) which represents the result of integrating the differential equation exactly through a unit interval. There is a trivial relationship between the group element for a method with step size h and the same method with unit step size. Hence, for convenience, we will always consider h to be equal to 1 in discussing algebraic properties of methods. For each $n = 1,2,3, \cdots$, there is a normal subgroup G_n such that $\alpha p^{-1} \in G_n$ if and only if the method, for which α is the corresponding member of G, is of order n.

To compute the elements of G, represented as a real valued function on the set of rooted-trees T, corresponding to a method m given by

$$\begin{array}{c|ccc} c_1 & a_{11} & a_{12} & \cdots \\ c_2 & a_{21} & a_{22} & \cdots \\ \vdots & \vdots & \vdots & \\ \hline & b_1 & b_2 & \cdots \end{array}$$

we define the function value for the tree with only a single node as $b_1 + b_2 + \cdots$ and generally, for (rooted) trees u,v the value will be of the form $b_1U_1+b_2U_2+ \cdots$, $b_1V_1+b_2V_2+ \cdots$ respectively where $U_1, U_2, \cdots, V_1,V_2, \cdots$ are functions of $a_{11},a_{12}, \cdots$ but not of $b_1, b_2, \cdots$. If uv denotes the tree formed by adjoining the roots of u and v but regarding the original root of u as the root of uv then we compute the value of the group element at uv as $\sum_i b_i U_i \sum_j a_{ij} V_j$. This procedure, which is described formally in [4] constitutes a recursive definition of the value of the group element for every rooted tree.

Members of G_n are such that, for each tree with no more than n nodes, the value of a member evaluated at that tree is zero. Thus, a method is of order n if for each tree with no more than n nodes, p and the corresponding group element take on the same values.

If α corresponds to the method m and β to the method m_1 in the definition of effective order, then we see that m is of effective order m if and only if $\beta^{-1}\alpha\beta\, p^{-1} \in G_n$ for some m_1. That is, if and only if for some β, $(\alpha\beta)(t) = (\beta p)(t)$ for every t with no more than n nodes. We will illustrate the detailed meaning of this concept in the case $n = 4$. First, however, we state, without proof, that in this criterion for effective order there is no loss in generality in supposing that $\beta(\tau) = 0$ for τ the tree with only one node.

Let $t_0 = \tau$, $t_1 = \tau\tau$, $t_2 = \tau\tau.\tau$, $t_3 = \tau.\tau\tau$, $t_4 = (\tau\tau.\tau)\tau$, $t_5 = \tau\tau.\tau\tau$, $t_6 = \tau(\tau\tau.\tau$ $t_7 = \tau(\tau.\tau\tau)$ be the eight trees of order less than 5, using the notation for the product of trees described above and more fully in [4]. We will write $\alpha_0 = \alpha(t_0)$,

$\alpha_1 = \alpha(t_1), \cdots, \beta_0 = \beta(t_0), \beta_1 = \beta(t_1) = \cdots, p_0 = p(t_0) = 1, p_1 = p(t_1) = \frac{1}{2}, \cdots$.

In the table below are shown, for each $t = t_0, t_1, \cdots, t_7$, the formula for $(\alpha\beta)(t)$ and a simplification of the equation $(\alpha\beta)(t) = (\beta p)(t)$. Throughout this table it is assumed that $\beta_0 = 0$.

t	$(\alpha\beta)(t)$	p(t)	$(\beta p)(t)$		$(\alpha\beta)(t) = (\beta p)(t)$
t_0	α_0	1	1	(3)	$\alpha_0=1$
t_1	$\alpha_1+\beta_1$	$\frac{1}{2}$	$\beta_1+\frac{1}{2}$	(4)	$\alpha_1=\frac{1}{2}$
t_2	$\alpha_2+2\alpha_0\beta_1+\beta_2$	$\frac{1}{3}$	$\beta_2+\frac{1}{3}$	(5)	$\alpha_2+2\beta_1=\frac{1}{3}$
t_3	$\alpha_3+\alpha_0\beta_1+\beta_3$	$\frac{1}{6}$	$\beta_3+\beta_1+\frac{1}{6}$	(6)	$\alpha_3=\frac{1}{6}$
t_4	$\alpha_4+3\alpha_0\beta_2+3\alpha_0^2\beta_1+\beta_4$	$\frac{1}{4}$	$\beta_4+\frac{1}{4}$	(7)	$\alpha_4+3\beta_2+3\beta_1=\frac{1}{4}$
t_5	$\alpha_5+\alpha_0\beta_2+\alpha_0\beta_3+\alpha_0^2\beta_1+\alpha_1\beta_1+\beta_5$	$\frac{1}{8}$	$\beta_5+\frac{1}{2}\beta_1+\frac{1}{8}$	(8)	$\alpha_5+\beta_2+\beta_3+\beta_1=\frac{1}{8}$
t_6	$\alpha_6+2\alpha_0\beta_3+\alpha_0^2\beta_1+\beta_6$	$\frac{1}{12}$	$\beta_6+\beta_2+\frac{1}{12}$	(9)	$\alpha_6-\beta_2+2\beta_3+\beta_1=\frac{1}{12}$
t_7	$\alpha_7+\alpha_0\beta_3+\alpha_1\beta_1+\beta_7$	$\frac{1}{24}$	$\beta_7+\frac{1}{2}\beta_1+\beta_3+\frac{1}{24}$	(10)	$\alpha_7=\frac{1}{24}$

From this table, we see that the method is of effective order 4 iff there are numbers $\beta_1, \beta_2, \beta_3$ such that (3), (4), $\cdots$, (10) are all satisfied. Eliminating β_1, β_2, β_3, from (5), (7), (8) and (9) we find

(11) $$\alpha_6 - 2\alpha_5 + \alpha_4 - \alpha_2 + \frac{1}{4} = 0$$

which, with (3), (4), (6) and (10) constitute the conditions on the coefficients of m for effective order 4 to hold. In the next section we look at the particular case of two stage methods.

4. Two stage methods of effective order 4.

For the method (2), let A denote the matrix

$$A = \begin{bmatrix} a_{11} & a_{12} \\ a_{21} & a_{22} \end{bmatrix}$$

and b^T the row vector $[b_1, b_2]$. Also $\mathbb{1}$ will denote the vector

$$\mathbb{1} = \begin{bmatrix} 1 \\ 1 \end{bmatrix}$$

Let $z^2 - Uz + V$ denote the characteristic polynomial of A and $z^2 - uz + v$ the polynomial with zeros c_1, c_2. We can now express $\alpha_2, \alpha_3, \cdots, \alpha_7$ in terms of α_0, α_1, u, v, U and V. We have

$$\alpha_2 = \sum_{i=1}^{2} b_i c_i^2 = \sum_{i=1}^{2} b_i (uc_i - v) = u\alpha_1 - v\alpha_0$$

$$\alpha_3 = b^T A^2 1 = b^T (UA-V) 1 = U\alpha_1 - V\alpha_0$$

and similarly,

$$\alpha_4 = (u^2-v)\alpha_1 - uv\alpha_0$$

$$\alpha_5 = (Uu-V)\alpha_1 - Uv\alpha_0$$

$$\alpha_6 = (uU-v)\alpha_1 - uV\alpha_0$$

$$\alpha_7 = (U^2-V)\alpha_1 - UV\alpha_0$$

Hence, with the values $\alpha_0 = 1$, $\alpha_1 = \frac{1}{2}$, (6) and (10) become

$$\frac{1}{2}U - V = \frac{1}{6}$$

$$\frac{1}{6}U - \frac{1}{2}V = \frac{1}{24}$$

so that $U = \frac{1}{2}$, $V = \frac{1}{12}$. We can now simplify (11) to

$$(u-1)(6v-3u+2) = 0$$

We distinguish two cases; in Case I, $u = 1$ and v is less than $\frac{1}{4}$ but otherwise arbitrary while in Case II, $6v - 3u + 2 = 0$.

5. Particular methods in Case I.

If $v = \frac{1}{4} - \theta^2$ and $\theta > 0$ we find the method to be

$$\begin{array}{c|cc} \frac{1}{2} - \theta & \frac{1}{4} - \frac{\theta}{2} + \frac{1}{24\theta} & \frac{1}{4} - \frac{\theta}{2} - \frac{1}{24\theta} \\ \frac{1}{2} + \theta & \frac{1}{4} + \frac{\theta}{2} + \frac{1}{24\theta} & \frac{1}{4} + \frac{\theta}{2} - \frac{1}{24\theta} \\ \hline & \frac{1}{2} & \frac{1}{2} \end{array}$$

We find that $\alpha_2 = \frac{1}{4} + \theta^2$, $\alpha_4 = \frac{1}{8} + \frac{3}{2}\theta^2$, $\alpha_5 = \frac{1}{12} + \frac{\theta^2}{2}$, $\alpha_6 = \frac{1}{24} + \frac{\theta^2}{2}$ so that β must satisfy $\beta_1 = \frac{1}{24} - \frac{\theta^2}{2}$, $\beta_2 = \beta_3 = 0$ to be consistent with (5), (7), (8) and (9). Let $p^{(c)}$ denote the group element corresponding to exact integration through a step size c, so that $p_0^{(c)} = c$, $p_1^{(c)} = c^2/2$, $p_2^{(c)} = c^3/3$, $p_3^{(c)} = c^3/6$, $p_4^{(c)} = c^4/4$, $p_5^{(c)} = c^4/8$, $p_6^{(c)} = c^4/12$, $p_7^{(c)} = c^4/24$. For the finishing formula we take a method corresponding to $\beta p^{(c)}$ for some c. Let $\gamma = \beta p^{(c)}$ so that, with the values of $\beta_1, \beta_2, \beta_3$, that have been agreed upon we have

$\gamma_0 = c$, $\gamma_1 = \frac{c^2}{2} + \frac{1}{24} - \frac{\theta^2}{2}$, $\gamma_2 = \frac{c^3}{3}$, $\gamma_3 = \frac{c^3}{6} + c(\frac{1}{24} - \frac{\theta^2}{2})$. It will be convenient for the finishing formula m_1 to have the form

$$\begin{array}{c|cc} c_1 & a_{11} & a_{12} \\ c_2 & a_{21} & a_{22} \\ \hline & \bar{b}_1 & \bar{b}_2 \end{array}$$

where $c_1, c_2, a_{11}, a_{12}, a_{21}, a_{22}$ are the same as for m. Since $\gamma_0 = \bar{b}_1 + \bar{b}_2$, $\gamma_1 = \bar{b}_1 c_1 + \bar{b}_2 c_2$ we have the following equations for $\bar{b}_1, \bar{b}_2$

$$\bar{b}_1 + \bar{b}_2 = c$$

$$\bar{b}_1 c_1 + \bar{b}_2 c_2 = \frac{c^2}{2} + \frac{1}{24} - \frac{\theta^2}{2}$$

while the requirements $\gamma_2 - u\gamma_1 + v\gamma_0 = 0$ and $\gamma_3 - U\gamma_1 + V\gamma_0 = 0$ each lead to the same restriction on c, that its value must be $\frac{1}{2}$ or $\frac{1}{2} + \theta\sqrt{3}$. In the case $c = \frac{1}{2}$, we find $\bar{b}_1 = \frac{1}{4} + \frac{\theta}{4} + \frac{1}{24\theta}$, $\bar{b}_2 = \frac{1}{4} - \frac{\theta}{4} - \frac{1}{24\theta}$. In the case $c = \frac{1}{2} + \theta\sqrt{3}$, we find $\bar{b}_1 = \frac{1}{4} + \frac{\sqrt{3}-1}{2}\theta + \frac{1}{24\theta}$, $\bar{b}_2 = \frac{1}{4} + \frac{\sqrt{3}+1}{2}\theta - \frac{1}{24\theta}$.

Rather than using m_1^{-1} as a starting formula, it is convenient to use $m_1^{-1}m$ so that this formula, besides introducing a perturbation to the starting value, moves it forward by a single step. This combined formula is equivalent to

$$\begin{array}{c|cc} c_1 - c & a_{11} - \bar{b}_1 & a_{12} - \bar{b}_2 \\ c_2 - c & a_{21} - \bar{b}_1 & a_{22} - \bar{b}_2 \\ \hline & b_1 - \bar{b}_1 & b_2 - \bar{b}_2 \end{array}$$

and becomes when $c = \frac{1}{2}$

$$\begin{array}{c|cc} -\theta & \frac{-3\theta}{4} & -\frac{\theta}{4} \\ \theta & \frac{\theta}{4} & \frac{3\theta}{4} \\ \hline & \frac{1}{4} - \frac{\theta}{4} - \frac{1}{24\theta} & \frac{1}{4} + \frac{\theta}{4} + \frac{1}{24\theta} \end{array}$$

or, when $c = \frac{1}{2} + \theta\sqrt{3}$

$$\begin{array}{c|cc} -\theta(1+\sqrt{3}) & -\frac{\sqrt{3}}{2}\theta & -\frac{(\sqrt{3}+2)}{2}\theta \\ \theta(1-\sqrt{3}) & \frac{2-\sqrt{3}}{2}\theta & -\frac{\sqrt{3}}{2}\theta \\ \hline & \frac{1}{4} - \frac{\sqrt{3}-1}{2}\theta - \frac{1}{24\theta} & \frac{1}{4} - \frac{\sqrt{3}+1}{2}\theta + \frac{1}{24\theta} \end{array}$$

6. Particular methods in Case II

The condition $6v - 3u + 2 = 0$ is equivalent to $(c_1-\frac{1}{2})(c_2-\frac{1}{2}) + \frac{1}{12} = 0$ so that if $c_1 = \frac{1}{2} - \theta$ then $c_2 = \frac{1}{2} + \frac{1}{12\theta}$. The method m becomes

$$\begin{array}{c|cc} \frac{1}{2}-\theta & \frac{1}{2(1+12\theta^2)} & \frac{-\theta(1-6\theta+12\theta^2)}{1+12\theta^2} \\ \frac{1}{2}+\frac{1}{12\theta} & \frac{1+6\theta+12\theta^2}{12\theta(1+12\theta^2)} & \frac{6\theta^2}{1+12\theta^2} \\ \hline & \frac{1}{1+12\theta^2} & \frac{12\theta^2}{1+12\theta^2} \end{array}$$

Note that if $\theta = \sqrt{3}/6$, case II becomes identical to case I with this same value of θ and is in fact the method with order 4 (in the usual sense). Excluding this value of θ, we approach the problem of finding a finishing formula as for case I. It turns out that $c = (3+\sqrt{3})/6$ and $\theta = (\sqrt{3}+\sqrt{15})/12$ (or one of the three other conjugate cases where the sign of one or both of $\sqrt{3}$ and $\sqrt{15}$ is changed in c and θ). These values lead to $\bar{b}_1 = \frac{3+\sqrt{3}}{12} - \frac{(\sqrt{3}-1)}{4\sqrt{15}}$, $\bar{b}_2 = \frac{3+\sqrt{3}}{12} + \frac{\sqrt{3}-1}{4\sqrt{15}}$.

Apart from noting that Case II methods are of order 3 (in the usual sense) whereas Case I methods are in general of order only 2, there has not seemed any pressing reason for studying their properties in detail. However, much of the rest of this paper applies to methods of both classes.

7. Region of absolute stability.

When the differential equation $y'(x) = q\,y(x)$ is solved by method (1), y_N and y_{N-1} are related by

$$y_N = R(hq)\,y_{N-1}$$

where R is a rational function of degrees (2,2). Let

$$R(z) = \frac{e_0+e_1z+e_2z^2}{1+d_1z+d_2z^2}$$

then, it is easy to see that

$$(e_0+e_1z+e_2z^2) - (1+d_1z+d^2)(1+\alpha_0z+\alpha_1z^2+\alpha_3z^3+\alpha_7z^4) = O(z^5)$$

as $z \to 0$.

From (3), (4), (6), (10) we see that if the method is to have effective order 4, e_0, e_1, e_2, d_1, d_2 have the same conditions imposed upon them as if the method were of order 4 (in the usual sense). Thus, $\phi(z)$ is the Padé approximation to e^z

$$R(z) = \frac{1+\frac{z}{2}+\frac{z^2}{12}}{1-\frac{z}{2}+\frac{z^2}{12}}$$

and, accordingly, the region of absolute stability (the region where $|R(z)| < 1$) is the open negative half plane.

8. An extrapolation property.

Let m, m_1 be the two methods introduced in the study of case I. If we calculate the coefficients in the methods m^{-1} and m_1^{-1} we find that, except for changes of sign and the ordering of the two stages, that these are the same as m and $m_1^{-1}m$ respectively if $c = \frac{1}{2}$ is used. Thus, in this case,

$$\phi(m^{-1},h,f) = \phi(m,-h,f),$$

$$\phi(m_1^{-1},h,f) = \phi(m_1^{-1}m,-h,f).$$

We now compare the solution computed at the point $x_0 + Nh$ using step size h with that computed using step size $-h$. The two results are

$$(\phi(m_1,h,f) \circ \phi(m,h,f)^{N-1} \circ \phi(m_1^{-1}m,h,f))\ (y_0)$$

and

$$(\phi(m_1,-h,f) \circ \phi(m,-h,f)^{-N-1} \circ \phi(m_1^{-1}m,-h,f))\ (y_0).$$

For sufficiently small $|h|$, we have

$$\begin{aligned}
&\phi(m_1,-h,f) \circ \phi(m,-h,f)^{-N-1} \circ \phi(m_1^{-1}m,-h,f) \\
&= \phi((m_1^{-1}m)^{-1},h,f) \circ \phi(m^{-1},-h,f)^{N+1} \circ \phi(m_1^{-1},h,f) \\
&= \phi(m^{-1}m_1,h,f) \circ \phi(m,h,f)^{N+1} \circ \phi(m_1^{-1},h,f) \\
&= \phi(m_1,h,f) \circ \phi(m,h,f)^{-1} \circ \phi(m,h,f)^{N} \circ \phi(m_1^{-1}m,h,f) \\
&= \phi(m_1,h,f) \circ \phi(m,h,f)^{N-1} \circ \phi(m_1^{-1}m,h,f)
\end{aligned}$$

so that the two computed results are identical and, accordingly the global truncation error is an even function of h. If appropriate smoothness conditions hold on f to allow an asymptotic expansion of the error, then this expansion contains only terms of even degree. Thus, methods of Case I can be made the basis of h^2-extrapolation methods.

9. A multistep formulation.

In the method (1) where $a_{11}, a_{12}, \cdots$ are as for Case I, let ξ_{N-1}, η_{N-1} be defined as

$$\xi_{N-1} = y_{N-1} + a_{11}k_1 + a_{12}k_2$$

$$\eta_{N-1} = y_{N-1} + a_{21}k_1 + a_{22}k_2$$

respectively, and using $\bar{b}_1, \bar{b}_2$ from the method m_1, let

$$z_{N-1} = y_{N-1} + \overline{b}_1 k_1 + \overline{b}_2 k_2$$

so that $z_0, z_1, \cdots$ correspond to (4th order) accurate results at points spaced with step size h. We can now formulate the algorithm for computing ξ_0, η_0 (starting values) and ξ_N, η_N, z_N for $N = 1,2,\cdots$ as follows, where it is supposed that the differential equation is written in autonomous form

$$\xi_0 = z_0 + h((a_{11}-\overline{b}_1)f(\xi_0) + (a_{12}-\overline{b}_2)f(\eta_0))$$

$$\eta_0 = z_0 + h((a_{21}-\overline{b}_1)f(\xi_0) + (a_{22}-\overline{b}_2)f(\eta_0))$$

$$\xi_N = z_{N-1} + h((b_1-\overline{b}_1)f(\xi_{N-1}) + (b_2-\overline{b}_2)f(\eta_{N-1}) + a_{11}f(\xi_N) + a_{12}f(\eta_N))$$

$$\eta_N = z_{N-1} + h((b_1-\overline{b}_1)f(\xi_{N-1}) + (b_2-\overline{b}_2)f(\eta_{N-1}) + a_{21}\, f(\xi_N) + a_{22}f(\eta_N))$$

$$z_N = z_{N-1} + h((b_1-\overline{b}_1)f(\xi_{N-1}) + (b_2-\overline{b}_2)f(\eta_{N-1}) + \overline{b}_1 f(\xi_N) + \overline{b}_2 f(\eta_N))$$

and substituting the values for Case I with $c = \frac{1}{2}$, we have

$$\xi_0 = z_0 + h\left[\frac{-3\theta}{4}f(\xi_0) + \frac{-\theta}{4}f(\eta_0)\right]$$

$$\eta_0 = z_0 + h\left[\frac{\theta}{4}\, f(\xi_0) + \frac{3\theta}{4}\, f(\eta_0)\right]$$

$$\xi_N = z_{N-1} + h\left[(\tfrac{1}{4}-\tfrac{\theta}{4}-\tfrac{1}{24\theta})f(\xi_{N-1}) + (\tfrac{1}{4}+\tfrac{\theta}{4}+\tfrac{1}{24\theta})f(\eta_{N-1}) + (\tfrac{1}{4}-\tfrac{\theta}{2}+\tfrac{1}{24\theta})f(\xi_N) + (\tfrac{1}{4}-\tfrac{\theta}{2}-\tfrac{1}{24\theta})f(\eta_N)\right]$$

$$\eta_N = z_{N-1} + h\left[(\tfrac{1}{4}-\tfrac{\theta}{4}-\tfrac{1}{24\theta})f(\xi_{N-1}) + (\tfrac{1}{4}+\tfrac{\theta}{4}+\tfrac{1}{24\theta})f(\eta_{N-1}) + (\tfrac{1}{4}+\tfrac{\theta}{2}+\tfrac{1}{24\theta})f(\xi_N) + (\tfrac{1}{4}+\tfrac{\theta}{2}-\tfrac{1}{24\theta})f(\eta_N)\right]$$

$$z_N = z_{N-1} + h\left[(\tfrac{1}{4}-\tfrac{\theta}{4}-\tfrac{1}{24\theta})f(\xi_{N-1}) + (\tfrac{1}{4}+\tfrac{\theta}{4}+\tfrac{1}{24\theta})f(\eta_{N-1}) + (\tfrac{1}{4}+\tfrac{\theta}{4}+\tfrac{1}{24\theta})f(\xi_N) + (\tfrac{1}{4}-\tfrac{\theta}{4}-\tfrac{1}{24\theta})f(\eta_N)\right]$$

References

1. J.C. Butcher, "Implicit Runge-Kutta processes", *Math. Comp.* 18 (1964), 50-64.

2. J.C. Butcher, "The effective order of Runge-Kutta methods", *Conference on the Numerical Solution of Differential Equations*, (Lecture Notes in Mathematics 109), Springer-Verlag (1969), 133-139.

3. J.C. Butcher, "The order of numerical methods for ordinary differential equations", *Math. Comp.* 27 (1973), 793-806.

4. J.C. Butcher, "An algebraic theory of integration methods", *Math. Comp.* 26 (1972), 79-106.

AN OVERVIEW OF SOFTWARE DEVELOPMENT FOR SPECIAL FUNCTIONS*

W. J. Cody

1. Introduction

There are three distinct steps in the development of a numerical computer program: the development of theoretical methods to perform the desired computation, the development of practical computational algorithms utilizing one or more theoretical methods, and the implementation of these practical algorithms in documented computer software. This paper concentrates on the third step from the viewpoint of a numerical analyst working on software for elementary and special functions.

In the case of special functions we normally think of the development of theoretical methods as the determination of various representations such as analytic expansions, both convergent and asymptotic, and minimax approximations. But there are many other theoretical approaches exemplified by Newton iteration for fractional powers, the arithmetic-geometric mean applied to the computation of elliptic integrals and recurrence methods for certain Bessel functions. Gautschi has recently written a superb survey [7] of this type of activity which we commend to the interested reader.

The second level of activity is the synthesis of practical computational algorithms based on the theoretical work. Such algorithms frequently combine several theoretical methods, each method restricted to that set of parameters for which it performs best, with a description of the decision processes and auxiliary computations necessary to link the various methods together. The preparation of an algorithm requires a gross knowledge of computer characteristics and a feeling for the ultimate design of software implementing the algorithm.

As an example of the difference between these first two levels of activity, Clenshaw's tables of coefficients for Chebyshev series expansions for selected functions [2] are a product of the first level whereas the algorithms by Clenshaw, Miller

*Work performed under the auspices of the U.S. Energy Research and Development Administration.

and Woodger [3], and by Miller [11] based on Clenshaw's coefficients represent the second level. These algorithms are careful but unpolished recipes for numerically determining function values from arguments to be supplied. For example, Clenshaw represents the tangent as

$$\tan(\pi x/4) = x \,\Sigma' \, a_{2r} T_{2r}(x), \qquad |x| \leq 1 ,$$

where $T_k(x)$ is the Chebyshev polynomial of degree k in x

$$T_k(x) = \cos(k \arccos x)$$

and the primed sum indicates that only one half of the first term is to be used. Since the tangent is defined for almost all real arguments ω, but the expansion involves $\pi x/4$ for $|x| \leq 1$, Miller's algorithm [11] includes a scheme for reducing ω to an appropriate x. It also includes tests to detect arguments that are too extreme to be meaningfully processed, but does not numerically specify threshold parameters for extreme arguments since these parameters are computer dependent. This is one way in which the algorithm is an unpolished recipe.

The algorithm produced at this second level of activity, even an algorithm such as Miller's presented in an algebraic computer language, is not a computer program. We want to emphasize the difference. A computer program exists only in a computer system. Miller's published algorithm for the tangent function is not an element of the NAG library [10], for example, but Schonfelder's subroutine S07AAF [10,13] which implements Miller's algorithm is.

There is a tendency among the computer public to identify the algorithm and the software, sometimes to the detriment of one or the other. While it is not often that a superb implementation will enhance the reputation of a poor algorithm, it is not unusual for a poor implementation to stain the reputation of a good algorithm. This is one reason more numerical analysts are becoming involved in the third stage of software development--the process of turning an algorithm into a running, documented computer program. To be effective in this work the numerical analyst must understand the accepted design goals for the software as well as the strengths and weaknesses of the computer system he is to exploit. These are the activities and considerations we discuss.

2. Reliability

The most visible attribute of good numerical software is reliability--the ability of a program to perform a well-defined calculation accurately and efficiently. Different implementations of a given algorithm can differ widely in reliability on a given problem in a given computer environment. For example, consider the problem of finding tan(11) in short precision arithmetic on an IBM 360. We use the following program fragment

```
W = 11.0
Y = TAN(W)
V = (W*1.66673)/1.66673
Z = TAN(V)
```

together with four subroutines for calculating the tangent. The first, which we will denote by M, is a straightforward implementation of Miller's algorithm as published. The second, denoted by S, is Schonfelder's MARK 4, 1974 NAG subroutine for short precision on IBM equipment [10]. The third, $\overline{S}$, is Schonfelder's revised program [13]. The last, denoted by C, is a minor modification of subroutine M to be described shortly. Table I compares decimal representations of the function values computed by these four subroutines with the "correct" function value obtained from the 23D values of the sine and cosine in [1]. The results displayed in this table should disturb unsophisticated computer users.

TABLE I

A comparison of four subprograms for the short precision circular tangent on an IBM 360

Computation \ Argument	11.0	$fl\left(\frac{11.0*1.66673}{1.66673}\right)$
Tables	-225.95085	-
Subroutine M	-226.13164	-226.59224
Subroutine S	-225.97859	-226.43837
Subroutine $\overline{S}$	-225.74219	-226.18042
Subroutine C	-225.95082	-226.39006

First, let's examine the reasons for the differences between columns 1 and 2 of the table. There are two different sources of error in a function subroutine. *Transmitted error* is that error in the computed function value due to a small error in the argument. Let

$$y = f(\omega)$$

where $f(\omega)$ is a differentiable function, and let δy denote the relative error and Δy the absolute error in y. Then

$$\delta y = \frac{\Delta y}{y} \doteq \frac{dy}{y} = \frac{f'(\omega)}{f(\omega)}\, d\omega \doteq \frac{f'(\omega)}{f(\omega)}\, \Delta\omega = \omega\, \frac{f'(\omega)}{f(\omega)}\, \delta\omega \ . \tag{2.1}$$

The transmitted error δy is a scaling of the *inherited error* $\delta\omega$ by the factor $\omega f'(\omega)/f(\omega)$. The second type of error, *generated error*, is the error generated within the subroutine. It includes the error due to the truncation of an essentially infinite process at some finite point, such as the truncation of an infinite series after n terms, for example, as well as error due to the round-off characteristics of the machine. In particular, it includes the error due to the inexact representation of constants.

Since the same subroutines were used to compute the function values in both columns of Table I, the differences between columns cannot reasonably be attributed to generated error. Therefore, they must be due to transmitted error, hence to inherited error. In simple words, the function arguments must be different for the two computations.

The inherited error is easily determined in this case. The IBM 360 uses a base 16 sign-magnitude arithmetic in which floating point numbers are represented as

$$s = \pm 16^e \cdot f ,$$

where e is an integer exponent and f is a binary fraction containing t bits normalized so that $1/16 \leq f < 1$. Table II indicates how the number of significant bits in f varies with the magnitude of f, a phenomenon known as wobbling word length. For the short precision mode t is 24.

TABLE II

Significance of Hexadecimal Fractions

f	Binary representation of f	No. of significant bits in f
$1/2 \leq f < 1$	.1xxx...	t
$1/4 \leq f < 1/2$	.01xx...	t-1
$1/8 \leq f < 1/4$	.001x...	t-2
$1/16 \leq f < 1/8$	.0001...	t-3

The original argument $\omega = 11.0$ has the full 24 significant bits, but the intermediate result

$$11*1.66673 = 18.33403$$

in line 3 of our program fragment contains only t-3, or 21, significant bits, of which the last is subject to roundoff error. We can therefore expect that the final argument V is correct to only about 20 significant bits.

We can verify that this is roughly the error seen by working backwards. From the last line of Table I the transmitted relative error is estimated at $1.944*10^{-3}$. From (2.1) and 10 place trigonometric tables we estimate the inherited error as

$$\delta V \doteq -7.82*10^{-7} .$$

This can be translated into units of the last bit position (ULPs) by

$$\frac{11\delta V}{16} \doteq -5.38*10^{-7} \doteq -9.0*2^{-24} = -9.0 \text{ ULP}$$

which clearly involves the last four bits of V, as predicted.

The differences between columns in Table I are then apparently due to inherited error from the manufactured argument V, something for which the individual subroutines cannot be held accountable. Column 1, however, corresponds to an error-free argument, hence the error seen must be generated error and must reflect the care taken in the individual implementations.

The subroutine $\overline{S}$ is a major perturbation of Miller's algorithm which performs very poorly on this example. We will not consider it any further in this discussion. The only essential difference between the remaining three subroutines is in the way the variable x required in the Chebyshev series is obtained from the input argument W. Basically, the argument reduction scheme involves the computation

$$x = W*(4/\pi)-N$$

where N is an appropriate integer. There is usually a loss of significance in this reduction due to cancellation of leading significant bits, but that is not the first source of error. Since

$$4/\pi \doteq 1.27$$

contains only 21 bits in its hexadecimal representation, the product $W*4/\pi$ can be expected to be correct to only 21 bits even before the subtraction. This error is only magnified in importance by the subsequent cancellation of leading bits. Schonfelder correctly anticipated this problem in subroutine S and replaced the original scheme by

$$X = W/(\pi/4)-N$$

where

$$\pi/4 \doteq .785$$

contains 24 significant bits in its hexadecimal representation. The effectiveness of this simple modification can be seen by comparing the results for subroutines M and S in Table I.

The remaining error in the argument reduction scheme is probably a matter of philosophy. Since the argument W usually contains inherited error in its low order bits, and any cancellation of leading significant bits during argument reduction promotes that error to more important bit positions, there is apparently no reason to worry about what bit pattern is shifted into the low-order bits. This philosophy is correct as long as we do have an inherited error. In our example where there is no inherited error we must still pay the penalty for assuming that there is. In our opinion, high performance software should instead assume that there is no inherited error, and the reduced argument x should be calculated under the assumption that the given argument W is exact.

This is not difficult to do. Simply break W and $4/\pi$ each into two parts

$$W = W_1+W_2$$
$$4/\pi = C_1+C_2$$

where the second part is much smaller in magnitude than the first. Note that since $4/\pi$ is a universal constant, C_1 and C_2 are known to any desired precision. In fact, one possibility for C_1 is

$$C_1 = 5215.0/4096.0$$

which can be evaluated exactly in most computers. Then the argument reduction

$$x = W*4/\pi - N$$

can be rewritten as

$$x = (W_1*C_1-N)+W_1*C_2+W_2*C_1+W_2*C_2$$

which correctly fills the low order bits of x after the cancellation of leading significant bits occurs. This is the scheme used in subroutine C.

The example we have used to illustrate the differences between these computational schemes was carefully chosen to magnify the various errors. For most arguments the differences would not be so dramatic, but there would be some difference. In a statistical sense, programs incorporating the philosophy and programming care exemplified by subroutines S and C are more accurate than naive subroutines such as M.

3. Robustness

The second attribute of good numerical software is *robustness*--the ability of a computer program to detect and gracefully recover from abnormal situations without involuntarily terminating the computer run. Robust software detects improper arguments before using them, for example, and anticipates and circumvents computational anomalies such as underflow or overflow.

We again turn to Schonfelder's work on the NAG library for an example. Subroutine S10ABF [10,12] evaluates the hyperbolic sine by the formulae

$$\sinh(x) = \begin{cases} x \sum_r{}' a_r T_r(2x^2-1), & |x| \le 1 \\ \dfrac{e^x - e^{-x}}{2}, & 1 < |x| \le R \end{cases}$$

with an error return for $|x| > R$. For $|x| < 1$, the argument needed for the Chebyshev series is evaluated by the Fortran statement

```
X2 = 2.0*(2.0*X*X-1.0) ,
```

where the extra factor of 2.0 relates to Clenshaw's algorithm for evaluating such a series [2]. This expression is not robust, for if the floating point range of the computer is $[10^{-d},10^{d}]$, then for every x in the interval $[10^{-d},10^{-d/2})$ the intermediate result X*X is too small, and underflow occurs. Most computers will properly replace that result with 0.0 and proceed after writing out an underflow message. Even though the computation proceeds to an accurate answer the underflow message raises nagging doubts in the user's mind regarding the validity of his results. He cannot judge the importance of that message without knowing more about the program than he probably does. The underflow will never occur, however, if X2 is determined by the program segment

```
X2 = -2.0
IF (ABS(X) .GT. EPS)  X2 = 2.0*(2.0*X*X-1.0)
```

where EPS is chosen so that

```
2.0*X*X-1.0 = -1.0
```

in the machine arithmetic whenever $|X| \leq$ EPS. The value of EPS is not critical provided underflow is avoided and accuracy in X2 is retained.

The parameter R in Schonfelder's implementation is chosen to avoid destructive overflow. However, the value of R chosen is a crude threshold to prevent the overflow of exp(x). Since

$$\sinh(x) < e^x$$

for large x, a more robust program would provide that precise threshold R which prevents the overflow of sinh(x), and would recast the computational algorithm to avoid overflow for those arguments below the threshold but beyond the point where the exponential function overflows. A discussion of this and other peculiarities associated with the hyperbolic sine can be found in [4].

The foregoing discussion underscores the fact that there are differences of opinion on design goals for function programs. Schonfelder has very clearly stated his design goals for the NAG function subroutines [12,13]. Similar statements of different goals exist for FUNPACK [5] and for the Fortran library of elementary functions on IBM equipment [9]. In each case the design goals are reasonable and appropriate for the intended use of the software. More importantly, each designer has been largely successful in achieving his stated goals.

4. A Constructive Example

Contrary to the impression we may have left, the achievement of reliability and robustness need not be an impediment to program development. Consideration of these qualities early in the design stage can actually contribute to the effort in the sense that the program design will involve accurate solutions of precise mathematical problems instead of ad hoc solutions to imprecise problems.

Consider the design of a program to evaluate the functions $\Gamma(x)$ and $\ell n\Gamma(x)$ for a real argument x. Under the assumption that the argument x is exact, the program is to produce an accurate function value whenever such a value exists, is representable in the computer, and can be obtained without excessive effort; and it is to provide an error exit in all other cases. The computation is to be free of underflow and overflow.

Consider the computation of $\Gamma(x)$ first. For $x > 0$ the recurrence relation

$$\Gamma(x+1) = x\Gamma(x) \tag{4.1}$$

can be used to reduce the computation to that for $\Gamma(x)$ over some suitable interval of unit length. Two obvious choices of interval exist: Clenshaw [2] provides a Chebyshev polynomial expansion of $\Gamma(x)$ for $1 \leq x \leq 2$, and Hart, et al. [8] give minimax approximations for the interval $2 \leq x \leq 3$. Aside from the efficiency of evaluation of the minimax forms there is little to choose between these two alternatives for most machines. However, since $.5 < \Gamma(x) \leq 1$ for $1 \leq x \leq 2$, the hexadecimal representation of the function over this interval contains no leading zero bits, while the representation of $\Gamma(x)$ for $2 \leq x \leq 3$, where $1 \leq \Gamma(x) \leq 2$, contains three leading zero bits. Clenshaw's approximation is therefore potentially more accurate on IBM machines. (Clenshaw's companion expansion for $1/\Gamma(x)$, $1 \leq x \leq 2$ [2], which converges more rapidly than the one for $\Gamma(x)$, also suffers from poor hexadecimal normalization and is therefore not as acceptable for IBM programs.)

As x becomes larger, repeated use of (4.1) becomes inefficient and roundoff error accumulates excessively. Since Hart, et al. [8] provide efficient minimax approximations to $\ell n\Gamma(x)$ for $x \geq 12$, $\Gamma(x)$ can be computed as

$$\Gamma(x) = \exp(\ell n\Gamma(x)), \qquad 12 \leq x < \text{XBIG} ,$$

where XBIG is the argument at which $\Gamma(x)$ becomes too large to be represented in the computer. If XMAX is the largest machine representable number, then XBIG satisfies the equation

$$\Gamma(\text{XBIG}) = \text{XMAX} .$$

XBIG is conveniently found by Newton iteration using standard asymptotic forms to evaluate $\Gamma(x)$ and/or $\psi(x) = \Gamma'(x)/\Gamma(x)$. Appropriate values of XMAX for several large scientific computers are given in Table III, along with estimates of XBIG.

TABLE III

Some Machine Dependent Parameters

Computer	IBM 370	CDC 7600	UNIVAC 1108
Arithmetic Precision	Long	Single	Double
t	56	48	60
XMAX	$16^{63}(1-16^{-14})$	$2^{1070}(1-2^{-48})$	$2^{1023}(1-2^{-60})$
XBIG	57.574	177.803	171.489
XMIN	16^{-65}	2^{-975}	2^{-1025}
XMININV	$16^{-63}(1+16^{-13})$	2^{-975}	$2^{-1023}(1+2^{-59})$

From (4.1), $\Gamma(x) \to 1/x$ as $|x| \to 0$. Let XMIN denote the smallest positive number representable on the machine, and let XMININV represent the smallest positive machine number whose inverse is also representable. Then $\Gamma(x)$ is not representable for $|x| <$ XMININV, and an error exit must be made. On some machines, where XMIN = XMININV,

$\Gamma(x)$ is computable for all small non-zero x (see Table III).

There is a small region, XMININV $\leq$ x < XSMALL for which

$$\Gamma(x) = \frac{1}{x} ,$$

to machine accuracy, and the related computation of $\Gamma(1+x)$ can be suppressed. If the machine representation of a number allows t bits for the normalized binary fraction, then

$$1+2^{-t} = 1$$

in the computer. Hence,

$$\Gamma(1+2^{-t}) = \Gamma(1) = 1 ,$$

and

$$\text{XSMALL} = 2^{-t} .$$

The reflection formula

$$\Gamma(x) = \frac{\pi}{\sin(\pi x)\ \Gamma(1-x)} , \qquad x < 0 , \tag{4.3}$$

reduces the computation for negative arguments to a related one for positive arguments. The evaluation of $\sin(\pi x)$ is critical here, since rounding error in forming πx appears as inherited error to the sine routine. Argument reduction within the sine routine then magnifies the importance of this error whenever $x < -1$. However, if we let

$$X = -x$$
$$z = [X]$$

and

$$y = X-z ,$$

where [X] denotes the integer part of X, then

$$\sin(\pi x) = (-1)^{z+1} \sin(\pi y)$$

minimizes the inherited error by accurately removing the integer part of x before introducing the rounding error in the multiplication by π.

This preliminary argument reduction apparently provides the opportunity for an easy test for singularities, since $y = 0$ iff $x = -n$ where n is an integer. Of course, it is necessary to examine the representation of $\Gamma(x)$ when $y \neq 0$. The case $x \to 0$ has already been treated, so assume

$$x = -n+\varepsilon , \qquad n > 0 .$$

Now n cannot be too large since $-x$ cannot be much larger than XBIG if $1/\Gamma(1-x)$ in (4.3) is not to underflow. Thus $|y| = |\varepsilon| \geq 2^{-t+1}n$ and $|\sin(\pi y)| \geq \pi 2^{-t}n$ in the machine. From (4.3)

$$\frac{\pi}{\Gamma(1-x)} \leq |\Gamma(x)| < \frac{2^t}{n\Gamma(1-x)} < \frac{2^{t+1}}{n} , \tag{4.4}$$

which shows that $|\Gamma(x)|$ will not overflow for $y \neq 0$. There is still a lower bound for x below which $\Gamma(x)$ may underflow. The determination and use of a precise bound is difficult, perhaps falling in the design goal category of requiring excessive effort, and some compromise may be prudent at this point. The problem is exemplified by the fact that $\Gamma(x)$ may be representable for $x = -n(1+2^{-t+1})$, but may underflow for the algebraically larger argument $x = -n+.5$. The cleanest, but not most precise, condition to use is to restrict x to $-x < \min(XNEG, XBIG-1)$ where XNEG satisfies the equation

$$\left|\frac{1}{\Gamma(1+XNEG)}\right| = XMIN .$$

Extension of the subroutine to evaluate $\ln\Gamma(x)$ is straightforward. An upper bound for x must be determined beyond which $\ln\Gamma(x)$ overflows. The computation is similar to the determination of XBIG. Hart et al. [8] provide useful minimax approximations for $x \geq 12$, and Cody et al. [5] provide minimax approximations for several intervals spanning $0 < x \leq 12$. These latter approximations retain relative accuracy near $x = 1,2$ where $\ln\Gamma(x)$ vanishes. For $x \to 0$, $\ln\Gamma(x) \to -\ln(x)$ which is always representable whenever x is. The only problem is to decide how to handle the case $x < 0$. Possibilities include an error return or the computation of $\ln|\Gamma(x)|$ whenever it is representable. We leave the completion of the subroutine design to the interested reader.

The implementation of this design is just as important as the design itself, for only the implementation provides numbers on a machine. Careless implementation can neutralize a careful design. Reliability and robustness are still properties of software, not of designs.

While the implementation of a design is not always as numerically interesting as the design work, the problems encountered are challenging and varied. Argument thresholds must be precisely determined according to the recipes in the design, and then verified, for example. It is not a trivial task to use a computer to determine overflow thresholds while avoiding overflow in the determination. Even when the subroutine is essentially working, there may still be skewed error distributions in the final function computations because of biased rounding in the arithmetic. Careful analysis is required to properly bias low order bits in appropriate approximation coefficients and thereby restore a modicum of symmetry to the error. Each implementation poses similar but distinct numerical challenges to the analyst who cares to become involved.

5. Conclusion

Although we have limited our discussion to software for the evaluation of functions, reliability and robustness are desirable properties of numerical software in general. Clearly the more robust and reliable a numerical program is, the more the implementers have considered machine design in their work, and the harder it is to transfer that work and bring it up to specifications on other machines. This is the reason we said earlier that the numerical analyst must understand the accepted design goals for an item of numerical software as well as the design of the computer to be exploited if he is to be effective in software production.

We do not believe the achievement of reliability and robustness in any item of numerical software is the result of applying numerical tricks beyond the ken of the average analyst or programmer. We prefer to think that the product is the natural result of providing the professional attention to software development that it deserves.

References

1. Abramowitz, M. and Stegun, I. A. (Eds), *Handbook of Mathematical Functions with Formulas, Graphs and Mathematical Tables*, Nat. Bur. Standards Appl. Math. Series, 55, U.S. Government Printing Office, Washington, D.C., 1964.

2. Clenshaw, C. W., *Chebyshev Series for Mathematical Functions*, National Physical Laboratory Mathematical Tables, 5, Her Majesty's Stationery Office, London, 1962.

3. Clenshaw, C. W., Miller, G. F., and Woodger, M., Handbook Series Special Functions - Algorithms for Special Functions I, *Num. Math.* 4, 1963, pp. 403-419.

4. Cody, W. J., Software for the Elementary Functions, in *Mathematical Software*, J. R. Rice (Ed.), Academic Press, N.Y. and London, 1971, pp. 171-186.

5. Cody, W. J., The FUNPACK Package of Special Function Subroutines, *ACM Trans. Math. Software*, 1, 1975, pp. 13-25.

6. Cody, W. J., and Hillstrom, K. E., Chebyshev Approximations for the Natural Logarithm of the Gamma Function, *Math. Comp.* 21, 1967, pp. 198-203.

7. Gautschi, W., Computational Methods in Special Functions - A Survey, to be published in the Proceedings of the MRC Advanced Seminar on Special Functions, Madison, Wisconsin, March 31 - April 2, 1975.

8. Hart, J. F., Cheney, E. W., Lawson, C. L., Maehly, H. J., Mesztenyi, C. K., Rice, J. R., Thacher, H. C., Jr. and Witzgall, C., *Computer Approximations*, Wiley, N.Y. and London, 1968.

9. Kuki, H., Mathematical Function Subprograms for Basic System Libraries - Objectives, Constraints, and Trade-offs, in *Mathematical Software*, J. R. Rice (Ed.), Academic Press, N.Y. and London, 1971, pp. 187-199.

10. NAG Library Manual Mark 4, Volume IV, NAG, Oxford, 1974.

11. Miller, G. F., Handbook Series Special Functions - Algorithms for Special Functions II, *Num. Math.*, 7, 1965, pp. 194-196.

12. Schonfelder, J. L., Special Functions in the NAG Library, in *Software for Numerical Mathematics*, D. J. Evans (Ed.), Academic Press, N.Y. and London, 1974, pp. 285-300.

13. Schonfelder, J. L., The Production of Special Function Routines for a Multimachine Library, private communication.

Approximation methods for expanding operators

L Collatz

Summary. An attempt is made in this report to give a very rough survey on expanding operators. The phenomenon of expanding operators T seems to appear very often. Some classical fixed point theorems cover cases with expanding operators; numerical examples for these are given. Furthermore, there exists a fixed point theorem of Krasnoselskii, which is applicable to nonlinear integral equations of Hammerstein-type under certain conditions: for this numerical examples are given. But usually it is not yet possible to get exact inclusion theorems for solutions u of $u = Tu$.

A general numerical procedure, working in the last mentioned cases also for not well posed problems, and problems with several solutions, is described and applied in concrete cases. It was not the intention of this paper to give the greatest possible generality but to illustrate the situation by many examples. It is hoped that more mathematicians than hitherto will deal with expanding operators and that there will be much success in this new field of research in the future.

§1 Introduction and numerical procedure

Many problems of numerical analysis with systems of linear or nonlinear equations, differential equations, integral equations and others can be written in the form

$$u = Tu \tag{1.1}$$

where u is a wanted element of a linear space R, a vector, a function, a system of functions etc., and T is a given linear or nonlinear operator, which maps a domain of definition $D \subset R$ into R. For the numerical calculation of a fixed point u of the operator T one often uses the iteration procedure

$$u_{n+1} = Tu_n \qquad (n = 0,1,2,\ldots) \tag{1.2}$$

starting with an element $u_o \in D$. Many papers deal with the case when the operator T is globally contractive in D or at least locally contractive in a neighbourhood S of a fixed point u. But in practical problems T is very often neither globally nor locally contractive. In this case the following numerical procedure can be successful: starting with a function $u_o(x,a_1,\ldots,a_p)$ which depends on the coordinates $x = \{x_1,\ldots,x_m\}$ and a parameter vector $a = \{a_1,\ldots,a_p\}$, one calculates u_1 by

$$u_1(x,a_1,\ldots,a_p) = T\,u_o\,, \tag{1.3}$$

and determines the parameters a_ρ so that

(1.4) $$\phi(a_\rho) = M(u_1-u_o) = \text{Min}$$

with a suitable measure M. Often one chooses

(1.5) $$M(u_1-u_o) = \|\Phi \cdot (u_1-u_o)\|$$

with a certain norm $\|\cdot\|$. The positive factor ϕ may be helpful for simplifying or improving the numerical calculations. In order to get more accurate numerical results one does not calculate further iterates u_2, u_3 from (1.1) (because the process is often diverging or at least unstable), but increases the number of parameters a_ρ.

Example. (The chosen examples are very simple and only to illustrate the methods; often one can also treat them with other methods; see for example Collatz-Krabs [73]). Consider the equation

(1.6) $$y' = \frac{dy}{dx} = x - \frac{1}{y}$$

with the condition

(1.7) $$\lim_{x\to+\infty} y(x) = 0 .$$

One asks for the unknown value $c = y(0)$.

If one starts with chosen values c_o for c and uses shooting methods, the procedure is unstable (see fig. 1). One writes the iteration procedure (1.2) in the form

(1.8) $$\frac{du_o}{dx} = x - \frac{1}{u_1(x)} .$$

It is convenient to use the defect

(1.9) $$D(u_o) = \frac{du_o}{dx} - x + \frac{1}{u_o}$$

for numerical calculation. Thus the use of the supremum norm $\|h\|_\infty = \sup_{x\varepsilon[0,\infty)} |h(x)|$ leads to the following measure $M(u_1-u_o)$ in (1.5):

(1.10) $$M(u_1-u_o) = \|\phi.(u_1-u_o)\|_\infty = \|D(u_o)\|_\infty$$

with $\phi = \frac{x-u_o'}{u_o} > 0$, under the assumption $u_o > 0$ in $[0,\infty)$.

Starting with the simplest term

$$u_o = \frac{1}{a_1+x} ,$$

one gets the smallest value of $\|D(u_0)\|_\infty = \|u_0^{-1}-x-u_0^2\|_\infty = \|a_1-u_0^2\|_\infty$ for $a_1 = 2^{-1/3} \approx$ 0.793

with $y(0) \approx \sqrt[3]{2} \approx 1.26$.

Better results would be obtained with more parameters,

e.g. taking $u_0 = \dfrac{a_3+x}{a_1+a_2x+x^2}$, one gets $a_1 = 1.3163$, $a_2 = 1.0756$, $a_3 = 1.4080$,

$$\|D(u_0)\|_\infty = 0.0917.$$

I thank Mrs S Böttger and Mr Wildhack for the numerical calculation.

§2 Different types of fixed point theorems

There are three classes of fixed point theorems in functional analysis for operator equations of the form (1.1) $u = Tu$ with the following assumptions:

I. The operator T is contractive in the whole domain D; there exists a constant $K < 1$ so that

(2.1) $\rho(Tf,Tg) \leq K\,\rho(f,g)$ holds for all $f,g \in D$, where ρ is a distance.

For this case one has the well known theorem for contractive mappings. The applicability can be enlarged slightly by adding some hypothesis and admitting $K = 1$.

II. There exists a subset $M \subset D$ which is mapped by T into itself:

$$TM \subset M. \tag{2.2}$$

Usually one supposes for M (or for TM) certain properties such as boundedness, convexity, closedness, compactness, but it is not required that T be locally contractive in the whole of M. Examples for the theorems of this class II are the fixed point theorems due to Brouwer, Miranda, Schauder etc. (see for example Smart [74]).

III. No bounded open subset $M \subset D$ is known with the property $TM \subset M$. In this case very little is known; Krasnoselskii [64] gave a theorem for a certain class of expanding operators which belongs to this class III (see §6).

The fixed point theorems of the class II admit inclusion theorems for solutions u of (1.1) even in certain cases where the operator is expansive at the considered fixed point u but maps a greater domain M into itself. Therefore the class II means an important progress in the applicability of fixed point theorems.

This may be illustrated by some simple examples.

I. Experience with the least squares method:

The operator (defined in the x-y plane)

$$(2.3) \qquad T\binom{x}{y} = \begin{pmatrix} 2(\sin y)-y^3+\frac{1}{4}x \\ 2(x-x^3)+\frac{1}{4}\cos y \end{pmatrix}$$

maps the square $D = \{(x,y) \; ; \; |x| \leq 1, \; |y| \leq 1\}$

into itself and has at least one fixed point $\hat{p} = (\hat{x},\hat{y})$ in D corresponding to Brouwer's theorem.

But $\hat{P}$ is an expanding fixed point; fig. 2 shows some points $P \in D$ with arrows pointing in the direction of the points TP. By geometrical interpolation one can get points (x,y) for which

$$(2.4) \qquad Q = (x-Tx)^2 + (y-Ty)^2$$

becomes smaller; or one takes six points in a triangle, fig. 3, and substitutes Q by a quadratic polynomial, interpolating Q in these six points,

$$(2.5) \qquad Q = a + bx' + cx'^2 + dy' + ey'^2 + fx'y'$$

with the minimum at $\tilde{x},\tilde{y}$:

$$(2.6) \qquad \tilde{x} = (df-2be)\Delta^{-1}, \; \tilde{y} = (bf-2cd)\Delta^{-1}, \; \Delta = 4ce - f^2 .$$

One gets very good results if one chooses six points very near to the fixed point.

Starting with $a_1 = -0.16$, $a_2 = -0.06$ one observes the strong influence of the mesh size h (I thanks Mrs S Böttger and Mr Wildhack for the numerical calculation):

h	$\tilde{x}$	$\tilde{y}$	Q
0.02	-0.15780	-0.05894	$1.133 \cdot 10^{-6}$
0.004	-0.158553	-0.059589	$4.813 \cdot 10^{-10}$
0.001	-0.1585673	-0.0596038	$7.263 \cdot 10^{-13}$

II. The applications of Brouwer's fixed point theorem to nonlinear vibrations with period p are well known. For the equation $\ddot{x} = f(t,x,\dot{x})$ one asks for a fixed point of the transformation $T(x(0),\dot{x}(0)) = (x(p),\dot{x}(p))$; in the phase plane $x,\dot{x}$ shown in fig. 4 f is supposed to be periodic in t with period p (Reissig, Samsone, Conti [69]). There may be stable and unstable circuits, and Brouwer's theorem gives in certain cases inclusion theorems also for unstable circuits.

III. The nonlinear integral equation of Hammerstein-type

(2.7) $$y(x) = Ty(x) \quad \text{with} \quad Tz(x) = \lambda \int_0^1 \frac{dt}{e^x + [z(t)]^2}$$

has for large positive values of λ expansive fixed points. The contraction mapping theorem works only for small λ, but the theory of monotonically decomposible operators works for every positive λ; with v_0, v_1, w_0, w_1, (see J. Schröder [56], Collatz [66], p.352, Bohl [74])

$$v_0 = 0,\ w_1 = Tv_0 = \lambda e^{-x},\ w_0 = \lambda,\ v_1 = Tw_0 = \frac{\lambda}{e^x + \lambda^2},\quad v_0 \leq v_1 \leq w_1 \leq w_0,$$

and we get for every $\lambda > 0$ the existence of at least one fixed point and the inclusion for a solution y, (see fig. 5), with $v_1(x) \leq y(x) \leq w_1(x)$ (for this phenonemon, see Collatz [71]).

§3 Expansive Operators occur frequently

I. The simplest example may be the "expansive" operator T for vectors $u = (x,y)$ in the plane:

(3.1) $$T\binom{x}{y} = \begin{pmatrix} x/2 \\ 2y \end{pmatrix}.$$

Fig. 6 illustrates the mapping. The operator T is contractive along the x-axis, the iteration procedure

(3.2) $$u_{n+1} = Tu_n \qquad (n = 0,1,\ldots)$$

converges to the unique fixed point $u = (0,0)$ if u_0 is a point of the x-axis, and diverges for all other points; the procedure is unstable even on the x-axis. The phenomenon of expansivity is closely connected with instability.

II. Let us consider the linear integral equation

(3.3) $$u(x) = f(x) + \lambda Tu(x) \quad \text{with} \quad Tz(x) = \int_B K(x,t)\, z(t)dt,$$

where $x = (x_1,\ldots,x_n)$, $t = (t_1,\ldots,t_n)$ are points in a given domain B of the n-dimensional point space R^n, and $f(x)$, $K(x,t)$ are given continuous functions on B, resp. $B \times B$, and λ is a given constant. Suppose the eigenvalue problem

$$k\,T\,v = v$$

has eigenvalues k_j and eigenfunctions v_j and one can develop f, u and $u_{(n)}$ with respect to the eigenfunctions v_j:

$$u_{(n)} = \sum_j c_{j,n} v_j \ , \ f = \sum_j f_j v_j \quad \text{with} \quad v_j = k_j T v_j \ .$$

Then $$c_{j,n+1} = f_j + \frac{\lambda}{k_j} c_{j,n} \quad \text{holds.}$$

The iteration procedure is unstable, if one has $|\lambda| < |k_j|$ for at least one k_j.

This shows that one has to expect instability very often.

§4 Examples with one and more solutions

The procedure of §1 was applied to many cases. We give some examples

I. Expansive fixed point for a boundary value problem with an ordinary differential equation:

(4.1) $$-y''(x) = (3+x)y(x), \quad y(0) = 0, \quad y(2) = 1.$$

II. Two solutions of a boundary value problem with a partial differential equation for a function $u(x,y)$

(4.2) $$-\Delta u = r^2 + u^2 \quad \text{in } B \ , \quad u = 0 \quad \text{on} \quad \partial B \ ,$$

where $r^2 = x^2 + y^2$, $\Delta u = \dfrac{\partial^2 u}{\partial x^2} + \dfrac{\partial^2 u}{\partial y^2}$, $B = \{(x,y) \ , \ r < 1\}$.

We choose as first approximation

$$u_o = a_1(1-r^2)$$

and get corresponding to (1.3) from

$$-\Delta u_1 = r^2 + u_o^2 \quad \text{in } B \ ; \quad u_1 = 0 \quad \text{on} \quad \partial B:$$

$$\begin{aligned} 144 u_1 &= 22\, a_1^2 + 9 - 36\, a_1^2 r^2 + 9(2a_1^2 - 1) r^4 - 4\, a_1^2 r^6 \\ &= (1-r^2)\, P \quad \text{with} \quad P = 22\, a_1^2 + 9 + (9 - 14 a_1^2)\, r^2 + 4 a_1^2 r^4 \ . \end{aligned}$$

The condition $u_1 = u_o$ for $r = 0$ gives

$$u_1(0,0) \approx a_1 = \left\{ \begin{matrix} 6.4823 \\ 0.0631 \end{matrix} \right\}$$

Better results can be expected for the next approximation

$$u_o = a_1(1-r^2) + a_2(1-r^4) \ .$$

III. Similarly the problem with the same notation as in the example before

(4.3) $$-\Delta u = r^2 + \lambda u^3 \quad \text{in } B \ , \quad u = 1 \quad \text{on} \quad \partial B \ ,$$

has 3 solutions for large values of λ , but only the solution with small values of u is stable.

IV. Nonlinear Volterra Equation

(4.4)
$$y = Ty \quad \text{with} \quad Tz(x) = 1 + \int_0^x e^{x^t} [z(t)]^2 dt \quad .$$

One gets with the approximate solution $w = e^{a_2 x}$ in the interval $0 \leq x \leq 0.4$ a defect Dw with $|Dw| \leq 0.05$ for $a_2 = 1.4226$.

§5 Not well posed problems

I. Initial value problem for an elliptic equation:

A very simple model in the theory of oceanography can be described in the following way: The highness of water may be represented by a function $u(x,y)$ with

(5.1)
$$\begin{cases} \Delta u = 0 \quad \text{in} \quad B \\ u = [\cos\phi]^2 \quad \text{on} \quad \Gamma , \quad \frac{\partial u}{\partial r} = 0 \quad \text{on} \quad \Gamma . \end{cases}$$

Here polar coordinates r,ϕ with $x = r\cos\phi$, $y = r\sin\phi$, (see fig. 7), are used, $B = \{(r,\phi),\ r < 1,\ x > \cos(\pi/4)\}$, $\Gamma = \{(r,\phi),\ r = 1,\ |\phi| \leq \pi/4\}$ where Γ may be a part of the coast of the ocean. $\partial u/\partial r = 0$ on Γ means that no water enters the continent. u may be observed on Γ and $u(\pi/4,y) = \hat{u}(y)$ is wanted.

This is a Cauchy-problem for an elliptic equation and therefore not well posed. But looking only for bounded solutions with bounded derivatives one may consider and solve this problem. By using an approximate function of the form

$$v(r,\phi) = \sum_{\nu=0}^{p} a_\nu r^\nu \cos(\nu\phi)$$

the method gives results which can be used numerically, (see Bredendiek and Collatz [75]).

II. Fredholm-Integral equation of first kind.

(5.2)
$$Tu(x) = 1 - \int_0^1 \frac{u(t)}{1+x+t}\, dt = 0$$

The approximate solution (I thank Mrs Böttger and Mr Wildhack for the numerical work)

$$v(x) = -0.11239 + 0.66419x - 0.26781x^2 - 0.31324x^3 - 0.08313x^4$$

has a defect

$$|Tv(x)| \leq 0.00584 \, .$$

Of course this is not an error bound for $u(x)$. If there are more solutions, one could ask for the solution with smallest norm. Then one has a problem of Simultan-approximation, (see Bredendiek [69]).

III. Many other not well posed problems have the described unstable behaviour. For instance, one looks for a solution $u(x,t)$ of the heat-conduction equation

$$\frac{\partial u}{\partial t} = \frac{\partial^2 u}{\partial x^2} \tag{5.3}$$

with the given data

$$\begin{aligned} u(x,0) &= f_1(x) \quad \text{for} \quad 0 < x < a \\ u(0,t) &= f_2(t) \quad \text{for} \quad t > 0 \\ \frac{\partial u}{\partial x}(0,t) &= f_3(t) \quad \text{for} \quad t > 0 \, , \end{aligned}$$

especially for the values of $u(a,t)$ for $t > 0$.

§6 A fixed point theorem of Krasnoselskii and Hammerstein's Equation

Not much has been done in obtaining exact error bounds for approximate solutions in cases of type III in §2, in which the operator T is expansive and not mapping any bounded open set into itself. There exists the Theorem due to Krasnoselskii [64]. Let the compact operator T map a cone K in a Banach space into itself, $TK \subset K$; suppose there exist r, R with $0 < r < R$ so that for every $\varepsilon > 0$ and (a) for every $x \in K$ with $0 < \|x\| \leq r$ we have $Ax - (1+\varepsilon)x \notin K$, (b) for every $x \in K$ with $\|x\| \geq R$ we have $x - Ax \notin K$, then there exists at least one point $u \in K$ with $u = Tu$ and $r \leq \|u\| \leq R$.

This theorem was applied by Sprekels [75] to integral equations of Hammerstein type (notations B,x,t as in §3, Example II)

$$u(x) = \int_B K(x,t) f(t,u(t))dt \, . \tag{6.1}$$

Let K be continuous on $B \times B$ (except $x=t$) and have values between finite positive bounds, and let f be continuous on $B \times R^1$; suppose that there exist $m_0 > 0$, $M_0 \geq m_0$, $\alpha < 0$, $\beta > 0$ and continuous functions $a(x) \geq 0$, $b(x) \geq 0$, $a \not\equiv 0$, $b \not\equiv 0$ with

$$0 \leq f(x,z) \leq a(x)\, z^{1+\alpha} \quad \text{for} \quad z \in [0, m_0]$$

$$f(x,z) \geq b(x)\, z^{1+\beta} \quad \text{for} \quad z \in [M_0, \infty).$$

Then there exists at least one continuous solution $\hat{u}$ of (6.1) with $\hat{u}(x) > 0$ in B.

Sprekels [75] gives inclusion theorems for $\hat{u}$; for this purpose one has to approximate the given kernel $K(x,t)$ by a product of functions of x and t : $K(x,t) \approx p(x)q(t)$.

The better this approximation (Flügge [75] deals with this question) the better are the error bounds. The measure for the goodness of the approximation is the constant c :

(6.2) $$V = \inf_{B\times B} \frac{K}{p\,q}\,, \quad W = \sup_{B\times B} \frac{K}{p\,q}\,, \quad c = \frac{W}{V}\,.$$

Then one has the inclusion theorem

(6.3) $$\frac{r}{c}\, p(x) \leq \hat{u}(x) \leq R\, p(x) \quad \text{with}$$

$$r = \operatorname{Min} \frac{m_o}{\operatorname{Max} p}\,, \ [W \int_B q(t)\, a(t)\, (p(t))^{1+\alpha} dt]^{-1/\alpha}$$

$$R = \operatorname{Max} \frac{M_o c}{\operatorname{Min} p}\,, \ [V. c^{-1-\beta} \int_B q(t)\, b(t)\, (p(t))^{1+\beta}\, dt]^{-1/\beta}\,.$$

§7 Error bounds in numerical examples

I. In this way a solution $u(x)$ of the nonlinear (Hammerstein) equation

(7.1) $$u(x) = \int_0^1 \sqrt{1+x+t}\ \ [u(t)]^4 dt$$

is included in the interval

$$0.7991\ \ z(x) \leq u(x) \leq 0.9682\ \ z(x) \quad \text{with}$$

$$z(x) = [0.74641 + 0.54641\ x]^{1/2}$$

I thank Dr Sprekels for this numerical example.

II. Equation

(7.2) $$u(x) = \int_0^1 (x+t+e^{xt})\ [u(t)]^2 dt\ .$$

Inclusion for the kernel: $ae^{b(x+t)} \leq K(x,t) \leq cae^{b(x+t)}$ for $0 \leq x \leq 1,\ 0 \leq t \leq 1$ with $a = 0.9632$, $b = 0.7306$, $c = 1.142$.

Inclusion for the solution u :

$$z(x) \leq u(x) \leq c\cdot z(x)$$

$$z(x) = \hat{a}e^{bx}, \quad \text{with } \hat{a} = 0.2505 \text{ and the same } b = 0.7306,\ c = 1.142.$$

The average has an error of at most 7%.
I thank Mrs Böttger and Mr Wildhack for the numerical calculation.

III. Boundary value problems:

(7.3) $$-y''(x) = (1+x^2)[y(x)]^p, \quad y(1)+3y'(1) = y(-1)-3y'(-1) = 0 ;$$

these problems were treated for $p = 2$ and $p = 4$.

Literature

Bohl, E. [74]:Monotonie, Lösbarkeit und Numerik bei Operator-gleichungen, Springer, Berlin 1974, 255 p.

Bredendiek, E. [69]:Simultanapproximation, Arch. Rat. Mech. Anal. 33, 307-330 (1969).

Bredendiek, E. and L. Collatz [75]:Simultanapproximation bei Randwertaufgaben, to appear in Internat. Ser. Num. Math., Birkhäuser, 1975.

Collatz, L. [66]:Functional Analysis and Numerical Analysis, Academic Press 1966, 473 p.

Collatz, L. [71]:Some applications of functional analysis, particularly to nonlinear integral equations, Proc. Nonlinear functional analysis and applications, Madison, Academic Press, 1971, 1-43.

Collatz, L. und W. Krabs [73]:Approximationstheorie, Teubner, Stuttgart 1973, 208 p.

Flügge, H. [75]:Zur Tschebyscheff-Approximation mit Funktionen mehrerer Variablen, Dissertation Universität Hamburg 1975, 71 p.

Krasnoselskii, M.A. [64]:Positive solutions of operator equations, P. Noordhoff, Groningen 1964.

Reissig, R., G. Sansone and R. Conti [69],:Nichtlineare Differentialgleichungen höherer Ordnung, Roma 1969, 738 p.

Schröder, J. [56]:Das Iterationsverfahren bei allgemeinem Abstandsbegriff, Math. Z. 66, 111-116 (1956).

Smart, D.R. [74]:Fixed point theorems, Cambridge Univ. Press (1974) 93 p.

Sprekels, J. [75]:Einige Existenz- und Einschließungssätze für Fixpunkte expandierender Integraloperatoren, erscheint demnächst.

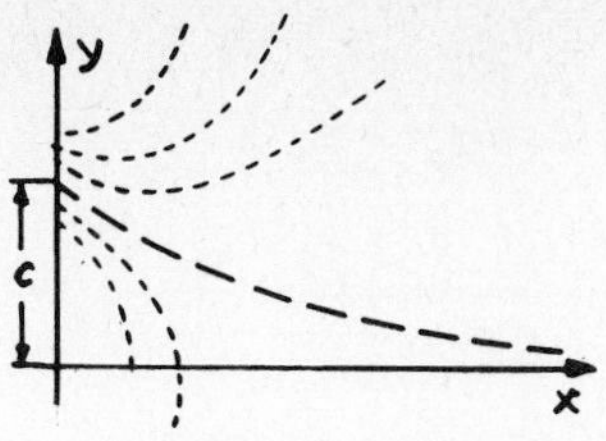

fig.1

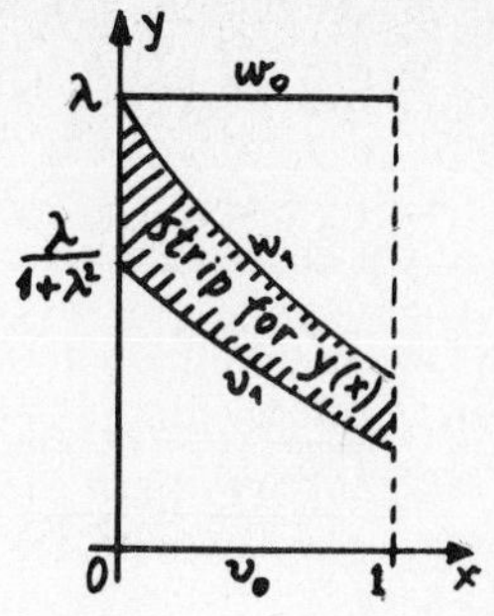

fig.5

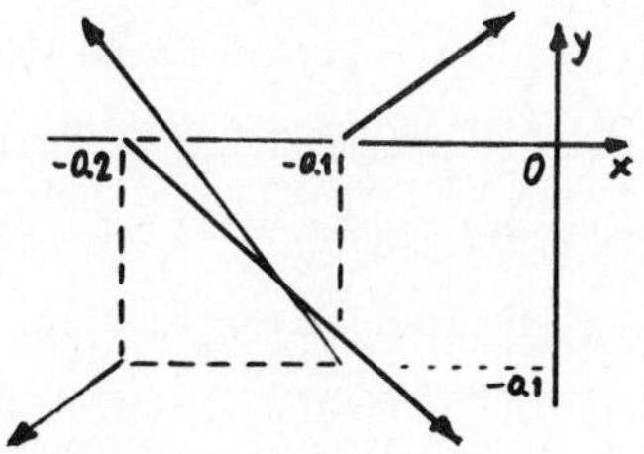

fig.2

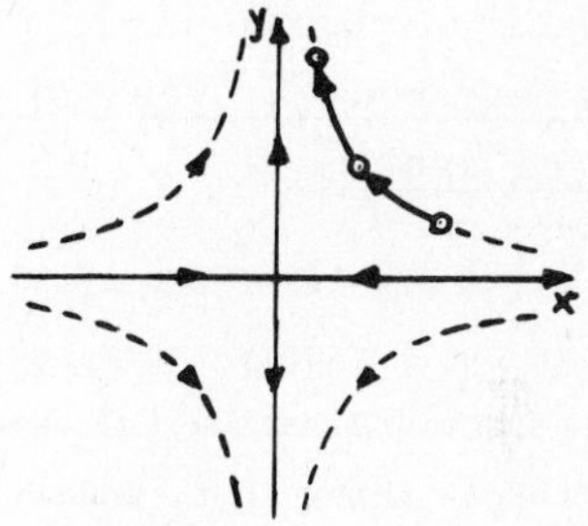

fig.6.

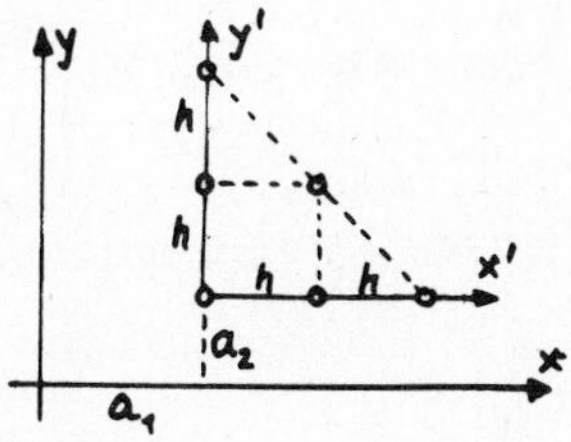

fig.3

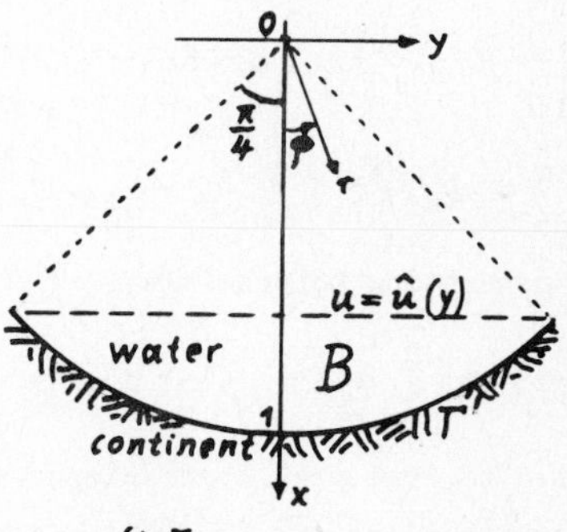

fig.7

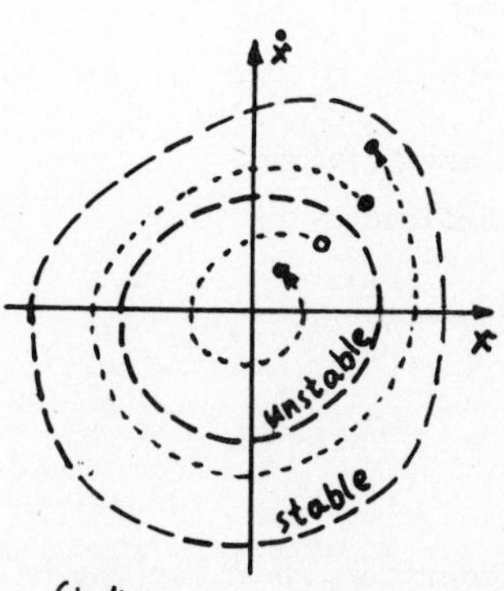

fig.4

ERROR ANALYSIS FOR A CLASS OF METHODS FOR STIFF NON-LINEAR INITIAL VALUE PROBLEMS

Germund Dahlquist

1. Preliminaries about a class of numerical methods and a class of differential systems

The theoretical analysis of the application of numerical methods on stiff non-linear problems is still fairly incomplete. The author knows of treatments of two methods: the trapezoidal method [2] and the backward Euler method [6]. The main purpose of the present paper is to modify and extend the approach in [2] to more general linear multistep methods and to a related class of methods, which will be called one-leg (multistep) methods.

Consider a system of ordinary differential equations,

$$\frac{dy}{dt} = f(t,y) \tag{1.1}$$

where f: $[a,b] \times R^s \to R^s$, f piece-wise analytic. A *linear k-step-method* is defined by the formula,

$$\sum_{j=0}^{k} \alpha_j \hat{y}_{n+j} = h \sum_{j=0}^{k} \beta_j f(t_{n+j}, \hat{y}_{n+j}) \tag{1.2}$$

where the generating polynomials,

$$\rho(\zeta) = \sum_{j=0}^{k} \alpha_j \zeta^j \qquad \sigma(\zeta) = \sum_{j=0}^{k} \beta_j \zeta^j , \tag{1.3}$$

are supposed to have real coefficients and no common divisor.

The corresponding *one-leg k-step method* is defined by,

$$\sum_{j=0}^{k} \alpha_j y_{n+j} = h\sigma(1) f\left(\frac{1}{\sigma(1)} \sum_{j=0}^{k} \beta_j t_{n+j}, \frac{1}{\sigma(1)} \sum_{j=0}^{k} \beta_j y_{n+j}\right) \tag{1.4}$$

We *assume* that

$$\sigma(1) = 1, \quad \beta_k > 0, \quad \alpha_k > 0 . \tag{1.5}$$

The first of these conditions is merely a convenient standardization, and the latter conditions are then satisfied for most methods for stiff problems, e.g. for all A_0-stable methods.

We shall use the term "*the method* (ρ,σ)" for statements which are valid for both the linear multistep method (1.2) and the one-leg multistep method (1.4).

When discussing the general non-linear case we assume, for the sake of simplicity, that it has been brought to *autonomous form*, $dy/dt = f(y)$, by augmenting the system with one more variable $y^0 = t$, and one more equation $dy^0/dt = 1$. The application of a consistent method to this system is equivalent to its application to the original non-autonomous system. The one-leg difference equation will therefore usually be written in the form,

$$\rho(E)y_n - hf\left(\sigma(E)y_n\right) = 0\,, \tag{1.4'}$$

where E *denotes the shift operator.*

As an example of this relationship, we consider the *trapezoidal method*, which is an implicit two-leg one-step-method,

$$y_{n+1} - y_n = \frac{h}{2}\left(f(y_{n+1}) + f(y_n)\right). \tag{1.6}$$

The corresponding one-leg method is the *implicit midpoint method*,

$$y_{n+1} - y_n = hf\left(\frac{y_{n+1} + y_n}{2}\right). \tag{1.7}$$

It turns out that the error analysis is simpler to formulate for one-leg methods than for linear multistep methods. Fortunately, a result for an one-leg method can be translated into a somewhat more complicated result for the corresponding linear multistep method, as a result of the following theorem. The proof is omitted, since it is an obvious modification of the proof for the perturbation-free case, given in [4].

THEOREM 1.1. *Let* $\{y_n\}$ *be a vector sequence which satisfies the (one-leg) difference equation* (1.4') *approximately, i.e.*

$$\rho(E)y_n - hf\left(\sigma(E)y_n\right) = p_n\,, \tag{1.8}$$

where $\{p_n\}$ *is a sequence of local perturbations. Put*

$$\hat{y}_n = \sum_{j=0}^{k} \beta_j y_{n+j} = \sigma(E)y_n\,. \tag{1.9}$$

Then $\{\hat{y}_n\}$ *satisfies the (linear multistep) difference equation*

$$\rho(E)\hat{y}_n - h\sigma(E)f(\hat{y}_n) = \sigma(E)p_n =: \hat{p}_n\,. \tag{1.10}$$

Conversely, let P,Q *be two polynomials of degree not exceeding* $k-1$, *such that for some integer* m, $0 \le m \le k$,

$$P(\zeta)\sigma(\zeta) - Q(\zeta)\rho(\zeta) \equiv \zeta^m\,. \tag{1.11}$$

Let $\{\hat{y}_n\}$ *be a vector sequence that satisfies* (1.10), *and put, for* $n \ge m$,

$$y_n = E^{-m}\left(P(E)\hat{y}_n - hQ(E)f(\hat{y}_n) - Q(E)p_n\right). \tag{1.12}$$

Then $\hat{y}_n = \sigma(E)y_n$, *and* $\{y_n\}$ *satisfies* (1.8), *for* $n \ge m$.

EXAMPLE 1.1. For the trapezoidal and the implicit midpoint methods,

$$\rho(\zeta) = \zeta - 1, \quad \sigma(\zeta) = \tfrac{1}{2}(\zeta+1)$$

Then, $\sigma(\zeta) - \frac{1}{2}\rho(\zeta) = 1$, i.e. $P(\zeta) = 1$, $Q(\zeta) = \frac{1}{2}$, if $m = 0$. Hence $\hat{y}_n = \frac{1}{2}(y_n + y_{n+1})$, $y_n = \hat{y}_n - \frac{1}{2}hf(\hat{y}_n)$. (In general, P and Q are found by Euclid's algorithm.)

EXAMPLE 1.2. The backward differentiation methods are one-leg methods, with $\sigma(\zeta) = \zeta^k$. If we choose $m = k$ we therefore have $P(\zeta) \equiv 1$, $Q(\zeta) \equiv 0$, $y_n = \hat{y}_{n-k}$ for $n \geq k$. If we choose $m = 0$ here, the polynomials become more complicated, see [4]. ∎

The error analysis for *one-leg methods* consists of estimating the difference between two sequences $\{y_n^*\}$ and $\{y_n^{**}\}$. The former is the sequence obtained in actual computation, which satisfies the equation

$$\rho(E)y_n^* - hf\left(\sigma(E)y_n^*\right) = p_n^*, \tag{1.13}$$

where p_n^* is a local perturbation, such as a rounding error or a truncation error in the iterative solution of the (algebraic) system which determines $\sigma(E)y_n^*$.

The latter sequence is defined by

$$y_n^{**} = y(t_n) + h^p e(t_n), \tag{1.14}$$

where $y(t)$ is the solution of the differential system, assumed to be a smooth function in $[a,b]$, i.e. the series $\Sigma y^{(n)}(t)(kh)^n/(n!)$ should converge rapidly. $e(t)$ is the dominant part of the global error, and should be a *smooth* solution of the variational equation,

$$de/dt = f'\left(y(t)\right)e - L(t), \tag{1.15}$$

$L(t)$ is defined by the equation

$$\rho(E)y(t) - hf\left(\sigma(E)y(t)\right) = h^{p+1}L(t) + \mathcal{O}(h^{p+2}). \tag{1.15'}$$

For a discussion of the existence of smooth solutions and their initial conditions, the reader is referred to [3], [8].

The sequence $\{y_n^{**}\}$ then satisfies a perturbed difference equation,

$$\rho(E)y_n^{**} - hf\left(\sigma(E)y_n^{**}\right) = p_n^{**} = \mathcal{O}(h^{p+2}). \tag{1.16}$$

Put

$$q_n = p_n^* - p_n^{**}, \quad z_n = y_n^* - y_n^{**}.$$

Then

$$\rho(E)z_n = h\left(f(\sigma(E)y_n^*) - f(\sigma(E)y_n^{**})\right) + q_n. \tag{1.17}$$

For *linear multistep methods* the vectors $\hat{y}_n^*$, $\hat{y}_n^{**}$, $\hat{p}_n^*$, $\hat{p}_n^{**}$, are similarly defined. Then, y_n^*, y_n^{**}, p_n^*, p_n^{**}, are defined as in Theorem 1.1, and we again obtain the error equation (1.17). We discuss one-leg methods only in the following.

The differential system is characterized by a choice of an inner-product $\langle\cdot,\cdot\rangle$

(and the related norm) in R^s, and a real constant μ, such that the one-sided Lipschitz condition,

$$\langle y-u,\ f(y)-f(u)\rangle \le \mu\,\|y-u\|^2\ ,\quad \forall y,\ u\in R^n \tag{1.18}$$

holds. Multiplying (1.17) by $\sigma(E)z_n$, we then obtain,

$$\langle \sigma(E)z_n,\ \rho(E)z_n\rangle \le h\mu\|\sigma(E)z_n\|^2 + \langle \sigma(E)z_n,\ q_n\rangle\ ,$$

and hence, for any $\eta > 0$,

$$\langle \sigma(E)z_n,\ \rho(E)z_n\rangle \le h(\mu + \tfrac{1}{2}\eta)\ \|\sigma(E)z_n\|^2 + \|q_n\|^2/(2\eta h) \tag{1.19}$$

We shall see in the following sections that, for many methods, it is possible to derive useful bounds for $\|z_n\|$ from this inequality.

LEMMA 1.2. *Suppose that* f: $R\times R^s \to R^s$ *is continuous and that* (1.18) *is satisfied and that*

$$h\mu < \alpha_k/\beta_k\ . \tag{1.20}$$

If $\sigma(E)t_n$, y_n, $y_{n+1}, \dots, y_{n+k-1}$ *are given, then* y_{n+k} *is uniquely determined by the equation,*

$$\rho(E)y_n = hf\big(\sigma(E)t_n,\ \sigma(E)y_n\big)\ .$$

PROOF: Put $\sigma(E)y_n = y$. The equation can then be written in the form,

$$y\alpha_k/\beta_k - hf(t,y) = x\ , \tag{1.21}$$

where $t\in R$, $x\in R^s$ are given. By (1.18) and (1.20),

$$\Big\langle y-u,\ (y-u)\,\alpha_k/\beta_k - h\big(f(t,y)-f(t,u)\big)\Big\rangle \ge \|y-u\|^2\,(\alpha_k/\beta_k - h\mu) \ge \delta\cdot\|y-u\|^2$$

for some $\delta > 0$, i.e. the left hand side of (1.21) is a uniformly monotone function, if the definition in [10, p.141] is harmlessly generalized to arbitrary inner-products in R^s. The lemma then follows from the uniform monotonicity theorem of Browder and Minty, [10, p.167]. ■

Note that (1.20) means *no restriction on* h, *if* $\mu < 0$.

In practice, the condition (1.18) is rarely valid in the whole space $R\times R^s$. This situation can be reduced to the case where the condition is valid in the whole space, because of the following lemma, which is based on ideas of B. Engquist and H.O. Kreiss (personal communication). A proof is given in [5].

LEMMA 1.3. *Let* p: $R\to R^s$, r: $R\to R$ *be given continuous functions, and let* f: $R\times R^s \to R$ *be a differentiable function such that* $f'_y(t,y)$ *is uniformly continuous in a tube in* $R\times R^s$, *defined by the conditions,*

$$t\in[a,b]\ ,\quad \|y-p(t)\| \le r(t)\cdot(1+\delta)\ ,\quad \textit{where } \delta > 0\ .$$

Suppose further that

$$\langle v, f'_y(t,y)v\rangle \le \mu\|v\|^2 ,$$

for all (t,y,v) *such that* $t\in[a,b]$, $v\in R^s$, $\|y-p(t)\|\le r(t)$.

Then there exists a function f^* *of the form,*

$$f^*(t,y) = \varphi\left(\left\|\frac{y-p(t)}{r(t)}\right\|^2\right) f(t,y) - \psi\left(\left\|\frac{y-p(t)}{r(t)}\right\|^2\right)\left(y-p(t)\right)$$

such that

$$f^*(t,y) = f(t,y) \quad \textit{when } \|y-p(t)\| \le r(t)$$

$$\langle v, f^*_y(t,y)\cdot v\rangle \le \mu\|v\|^2 , \quad \forall (t,y,v) \in [a,b]\times R^s\times R^s . \tag{1.22}$$

In fact, for any non-increasing differentiable function φ: $R\to R$ such that $\varphi(u)=1$ for $u\le 1$, $\varphi(u)=0$ for $u\ge(1+\delta)^2$, one can define a non-decreasing differentiable function ψ: $R\to R$, with $\psi(u)=0$ for $u\le 1$, such that (1.22) is valid. If p and r are piece-wise analytic, there is no difficulty in making f^* piece-wise analytic.

Let $p(t)$ be an exact solution of the differential system (1.1) and suppose that $r(t)$ can be chosen larger than the acceptable error of a numerical solution. Then $f^*(t,y) = f(t,y)$ in the interesting part of $R\times R^s$. We write f instead of f^* in the following, and assume the validity of (1.22) or, equivalently, (1.18).

The parameter μ can be looked upon as an upper bound for the logarithmic norm (sometimes called the measure and denoted by $\mu(A)$) of the Jacobian, corresponding to the inner-product norm. In fact,

$$\frac{\|I+\varepsilon A\|-1}{\varepsilon} = \sup_x \frac{\|x+\varepsilon Ax\|-\|x\|}{\varepsilon\|x\|} = \sup_x \frac{\|x+\varepsilon Ax\|^2-\|x\|^2}{\varepsilon\|x\|(\|x+\varepsilon Ax\|+\|x\|)}$$

and hence

$$\mu(A) = \lim_{\varepsilon\downarrow 0}\frac{\|I+\varepsilon A\|-1}{\varepsilon} = \sup_x \frac{\langle x,Ax\rangle}{\|x\|^2} . \tag{1.23}$$

Lemma 1.2 can be generalized to a general logarithmic norm, see Desoer and Haneda [6], but we have not been able to generalize Lemma 1.3 or the formalism of the following section. The general error estimate (see [1]) yields a useful upper bound for the solution of (1.15), $\|e(t)\|\le\gamma(t)$ where

$$\gamma' = \mu\gamma + \lambda(t) , \quad \gamma(a)\ge\|e(a)\| \tag{1.24}$$

where $\lambda(t)$ is some upper bound of $\|\underset{\cdot}{L}(t)\|$. (The reader can verify this by forming the inner-product of (1.15) and e, using the Schwarz inequality and (1.22), and then dividing by $\|e\|$.)

2. Error analysis for a class of methods

In this section, we consider an autonomous differential system, where f is piece-wise analytic and satisfies (1.18), and its numerical treatment by an one-leg method which satisfies (1.5). The stepsize h is constant in $[a,b]$. Our purpose is to obtain bounds for the vector sequence $z_n = y_n^* - y_n^{**}$, defined in Section 1. We recall the inequality (1.19),

$$\left\langle \sigma(E)z_n,\ \rho(E)z_n \right\rangle \le h(\mu + \tfrac{1}{2}\eta)\ \|\sigma(E)z_n\|^2 + \|q_n\|^2/(2\eta h)\ , \tag{2.1}$$

where $q_n = p_n^* - p_n^{**} = \mathcal{O}(h^{p+2})$ + local rounding error, and η is an arbitrary positive number, to be chosen later. Let $x_v \in R$, $z_v \in R^s$, and put

$$Z_n = \begin{bmatrix} z_{n+k-1} \\ z_{n+k-2} \\ \cdots \\ z_n \end{bmatrix}, \qquad X_n = \begin{bmatrix} x_{n+k-1} \\ x_{n+k-2} \\ \cdots \\ x_n \end{bmatrix}$$

DEFINITION: Let $G = [g_{ij}]$ be a real symmetric, positive definite matrix. The method (ρ,σ) is said to be G-*stable, iff for all real* $x_0, x_1, \dots, x_k$,

$$X_1^T G X_1 - X_0^T G X_0 \le 2\sigma(E)x_0 \cdot \rho(E)x_0\ . \tag{2.2}$$

This is equivalent to requiring that the matrix S,

$$S = \begin{bmatrix} \alpha_k \\ \cdots \\ \alpha_1 \\ \alpha_0 \end{bmatrix} [\beta_k, \dots, \beta_1, \beta_0] + \begin{bmatrix} \beta_k \\ \cdots \\ \beta_1 \\ \beta_0 \end{bmatrix} [\alpha_k, \dots, \alpha_1, \alpha_0] - \begin{bmatrix} G & 0_k \\ 0_k^T & 0 \end{bmatrix} + \begin{bmatrix} 0 & 0_k^T \\ 0_k & G \end{bmatrix} \tag{2.3}$$

be non-negative definite. (0_k denotes the k-dimensional zero column vector.)

We shall see in Section 3 that several interesting methods are G-stable for some G.

For the vector case, put

$$\mathrm{G}(Z_n) = \sum_{i=1}^{k} \sum_{j=1}^{k} g_{ij} \left\langle z_{n+k-i},\ z_{n+k-j} \right\rangle .$$

THEOREM 2.1. *If the method* (ρ,σ) *is* G-*stable then*

$$\mathrm{G}(Z_{n+1}) - \mathrm{G}(Z_n) \le 2\left\langle \sigma(E)z_n,\ \rho(E)z_n \right\rangle \tag{2.4}$$

for all vectors $z_n, z_{n+1}, \dots, z_{n+k}$.
This is a consequence (for $H = S$) of the following more general lemma.

LEMMA 2.2. *The function*

$$\sum_{i=1}^{r} \sum_{j=1}^{r} h_{ij} \left\langle z_{r-i},\ z_{r-j} \right\rangle , \qquad (h_{ij} = h_{ji})$$

is non-negative for all vectors $z_0, z_1, \dots, z_{r-1}$, *iff the matrix* $H = [h_{ij}]$ *is non-negative definite.*

PROOF: The "only if" part is obvious. For the "if"-part, note that we can find numbers $p_m \ge 0$, such that

$$\sum_{i=1}^{r}\sum_{j=1}^{r} h_{ij}x_{r-i}x_{r-j} = \sum_{m=1}^{r} p_m\left(\sum_{i=1}^{r} c_{mi}x_{r-i}\right)^2$$

Since the representation on the left hand side is unique for a quadratic form (when $h_{ij}=h_{ji}$) the same algebraic operations yield in the vector case,

$$\sum_{m=1}^{r} p_m \left\|\sum_{i=1}^{r} c_{mi}z_{r-i}\right\|^2 = \sum_{i=1}^{r}\sum_{j=1}^{r} h_{ij}\langle z_{r-i}, z_{r-j}\rangle ,$$

which proves the lemma. ■

Let β be a positive number and let b_σ be an upper bound of $\left(\sigma(E)x_0\right)^2$ subject to the constraint that $X_1^T GX_1 + X_0^T GX_0 = 1$. Then, by Lemma 2.2,

$$\|\sigma(E)z_n\|^2 \le b_\sigma\Big(\mathrm{G}(Z_{n+1}) + \beta\mathrm{G}(Z_n)\Big) \tag{2.5}$$

Insert this and (2.4) into (2.1). If $\mu+\frac{1}{2}\eta \ge 0$,

$$\mathrm{G}(Z_{n+1}) - \mathrm{G}(Z_n) \le h(2\mu+\eta)b_\sigma\Big(\mathrm{G}(Z_{n+1}) + \beta\,\mathrm{G}(Z_n)\Big) + \|q_n\|^2/(\eta h) \tag{2.6}$$

If $\mu < 0$, we obtain for $\eta=-2\mu$ the result,

$$\mathrm{G}(Z_{n+1}) \le \mathrm{G}(Z_0) + \sum_{v=0}^{n}\|q_v\|^2/|2\mu h| ,\qquad (\mu<0). \tag{2.7}$$

Bounds for $\|z_{n+k}\|$ are easily derived from this. Let θ^2 be an upper bound of x_k^2 when $X_1^T GX_1 = 1$. Then

$$\|z_{n+k}\|^2 \le \theta^2\,\mathrm{G}(Z_{n+1}) . \tag{2.8}$$

For example, if $\|q_n\| \approx \|p_n^{**}\| \le Ch^{p+2}$ and if the initial error is negligible, then

$$\|z_{n+k}\| \le \theta Ch^{p+1}\left|\tfrac{1}{2}(t_{n+k}-t_0)/\mu\right|^{1/2} .$$

Note that $\|z_{n+k}\| = \mathcal{O}(h^{p+1})$ while the dominant part of the error is $h^p e(t_{n+k})$. Also note that it is possible to obtain an error bound which is uniform with respect to t_n if $y(t)$ for increasing t is such that $\sum_0^\infty \|q_v\|^2$ is convergent.

By (2.6), provided that $h(2\mu+\eta)b_\sigma < 1$,

$$\mathrm{G}(Z_{n+1}) \le \left(\mathrm{G}(Z_0) + \frac{1}{\eta h}\sum_{v=0}^{n}\|q_v\|^2\right)\cdot\left(\frac{1+\beta h(2\mu+\eta)b_\sigma}{1-h(2\mu+\eta)b_\sigma}\right)^{n+1} .$$

The choice $\eta=-2\mu$ is not possible for positive μ. A better choice is then to put $\eta=\eta_n$, where

$$1/\eta_n = (\beta+1)b_\sigma(t_{n+1}-t_0) . \tag{2.9}$$

Then

$$\mathrm{G}(Z_{n+1}) \lesssim \left(\mathrm{G}(Z_0) + \frac{1}{\eta_n h}\sum_{v=0}^{n}\|q_v\|^2\right)\exp(1+2\mu/\eta_n) . \tag{2.10}$$

This bound is to be used for $\mu \ge 0$. *It is also valid for negative* μ, *as long as*

$$1 + 2\mu b_\sigma(\beta + 1)(t_{n+1} - t_0) \geq 0\,, \tag{2.11}$$

and is then sharper than (2.7). *Otherwise* (2.7) *is better for negative* μ.

3. Tests for G-stability

THEOREM 3.1. *G-stability implies A-stability.*

PROOF: Since we deal only with real systems, the complex test equation $dy/dt = (\lambda_1 + i\lambda_2)y$ is written as a real system, $dz/dt = \Lambda z$, where

$$\Lambda = \begin{bmatrix} \lambda_1 & -\lambda_2 \\ \lambda_2 & \lambda_1 \end{bmatrix}.$$

The difference equation reads,

$$\rho(E)z_n = h\Lambda\sigma(E)z_n$$

and by (2.4), for $n = 0,1,2,\ldots$, we obtain for the ℓ_2-norm,

$$G(Z_{n+1}) - G(Z_n) \leq 2\left\langle \sigma(E)z_n, h\Lambda\sigma(E)z_n \right\rangle = 2h\lambda_1 \|\sigma(E)z_n\|^2$$

Therefore, if $\lambda_1 \leq 0$ then $G(Z_{n+1}) \leq G(Z_n) \leq \ldots \leq G(Z_0)$, and hence (by (2.8)), $\|z_{n+1}\|$, $n = 0,1,2,\ldots$, is bounded, which proves A-stability. ∎

The search for a matrix G, or the proof of its non-existence, is simplified by the following necessary conditions.

THEOREM 3.2. *Suppose that the method* (ρ,σ) *is G-stable and that the order of accuracy is* p, $p \leq 2$. *Then*

$$(1,\ldots,1,1)G = (\alpha'_{k-1},\ldots,\alpha'_1,\alpha'_0) \tag{3.1}$$

where

$$\alpha'_k = 0,\quad \alpha'_{j-1} - \alpha'_j = \alpha_j,\quad j = k,k-1,\ldots,0\,. \tag{3.2}$$

In addition to this the following holds iff $p = 2$,

$$(k-1,\ldots,1,0)G = (\alpha''_{k-1},\ldots,\alpha''_1,\alpha''_0) \tag{3.3}$$

where

$$\alpha''_k = 0,\ \alpha''_{j-1} - \alpha''_j = \beta_j - \alpha'_{j-1} + \sigma'(1)\alpha_j\,. \tag{3.4}$$

As a result of these relations, the rank of the $(k+1)\times(k+1)$ *matrix* S *(defined by* (2.3)) *is at most* $k+1-p$ (even if the method is not known to be G-stable).

The details of the proof are given in [4]. The main idea in the proof of the first part is that, by (2.3),

$$(1,\ldots,1,1)S(1,\ldots,1,1)^T = 2\rho(1)\sigma(1) - \sum\sum g_{ij} + \sum\sum g_{ij} = 0\,,$$

for a consistent method. Since S is non-negative definite, it follows that

$$(1,\ldots,1,1)S = 0\,.$$

This together with (2.3) and (3.1) implies (3.2).

The two latter formulas are then equivalent to the equation

$$(k,\ldots\ldots,1,0)S = 0\,.$$

The proof is somewhat more intricate.

REMARK: The number of parameters in the symmetric matrix G is $k(k+1)/2$. One can show that the $2k$ linear relations in (3.1) and (3.3) are consistent (when $p=2$) and that $2k-1$ of them are linearly independent.

COROLLARY:
$$\sum_{i=1}^{k}\sum_{j=1}^{k} g_{ij} = 1 \tag{3.5}$$

By (3.1), (3.2), the consistency condition and (1.5),

$$\sum_{i=1}^{k}\sum_{j=1}^{k} g_{ij} = (1,\ldots,1,1)G(1,\ldots,1,1)^T = \sum_{v=0}^{k-1}\alpha_v' = \rho'(1) = \sigma(1) = 1\,.$$

THEOREM 3.3. *Every A-stable k-step method of order k (or more) is G-stable for exactly one matrix G.*

Sketch of proof: (Details are given in [4].) A-stable methods of order k or more exist only for $k=1$ and $k=2$. *For* $k=1$, the general form is

$$y_{n+1} - y_n = hf\Big((1-\theta)y_{n+1} + \theta y_n\Big), \quad \theta \le \frac{1}{2}.$$

By (3.1) the only possible "matrix" is $G=1$. We obtain G-stability, since by (2.4)

$$S = \begin{bmatrix} 1-2\theta & 2\theta-1 \\ 2\theta-1 & 1-2\theta \end{bmatrix}$$

which is clearly positive semidefinite, when $\theta \le 1/2$.

For $k=2$, $p=2$, the general A-stable formula is generated by

$$\begin{aligned} \rho(\zeta) &= \frac{1}{2}\Big((\zeta+1)(\zeta-1) + \gamma(\zeta-1)^2\Big) \\ \sigma(\zeta) &= \frac{1}{4}\Big((\zeta+1)^2 + \gamma(\zeta+1)(\zeta-1) + \delta(\zeta-1)^2\Big) \end{aligned} \tag{3.6}$$

where $\gamma \ge 0$, $\delta > 0$. The only possible symmetric 2×2 matrix G is obtained by solving the 3 independent relations (3.1) and (3.3) for the 3 parameters in G. The result is,

$$G = \frac{1}{4}\begin{bmatrix} (\gamma+1)^2+\delta & 1-\gamma^2-\delta \\ 1-\gamma^2-\delta & (\gamma-1)^2+\delta \end{bmatrix} \tag{3.7}$$

G is positive definite, because

$$g_{11} > 0,\, g_{11}g_{22} - g_{12}^2 = \frac{((\gamma+1)^2+\delta)((\gamma-1)^2+\delta) - (1-\gamma^2-\delta)^2}{16} = \frac{4\delta}{16} > 0\,.$$

It still remains to test whether S is positive semi-definite. To begin with, assume that γ is strictly positive. By Theorem 3.2, the rank equals 1 at most when $p=2$. It is therefore sufficient to find one positive diagonal element. In fact, by (2.4) and (3.6), the first diagonal element is

$$2\alpha_2\beta_2 - g_{11} = \frac{1}{4}(1+\gamma)(1+\gamma+\delta) - \frac{1}{4}\left((\gamma+1)^2 + \delta\right) = \frac{1}{4}\gamma\,\delta > 0 .$$

The case $\gamma = 0$ is settled because the set of non-negative definite matrices is closed. ∎

For $k=3$, $p=2$, G-stability is no longer equivalent to A-stability. J. Oppelstrup has found an example, [4].

It is nevertheless true, for all k, that *for any given polynomial* σ *which satisfies the root condition* (which is a *necessary* condition for A-stability) *it is possible to find a G and a* ρ, *such that the method* (ρ,σ) *becomes G-stable and consistent.*

For let B be the companion matrix of the polynomial $\sigma(\zeta)/\beta_k$ such that $\sigma(E)x_0 = 0 \Leftrightarrow X_1 = BX_0$, and put

$$K = \begin{bmatrix} 1 & & \\ 1 & & \\ \dots & B & \\ 1 & & \\ 1 & 0\ 0 \dots 0 & 1 \end{bmatrix} \qquad (\det K \neq 0)$$

It is shown in [4] that $K^T SK$ is of the form, (if $\rho(1) = 0$),

$$K^T SK = \begin{bmatrix} 0 & a^T \\ a & G - B^T GB \end{bmatrix} \tag{3.8}$$

Because of the root condition, a positive definite matrix G can be found so that $G-B^T GB$ is non-negative definite. (This is the one-matrix case of the matrix theorem of Kreiss.) A consistent method (ρ,σ) is then constructed my means of Eqs. (3.1) and (3.2). After some calculation it is seen that these imply $a = 0$, which implies that $K^T SK$ is positive semi-definite, and hence also that S is positive semidefinite, and this is equivalent to G-stability.

EXAMPLE. The second order *BDF* method,

$$\nabla y_{n+2} + \frac{1}{2}\nabla^2 y_{n+2} = h f_{n+2}$$

is obtained in (3.6) for $\delta = 1$, $\gamma = 2$. Then by (3.7)

$$G = \frac{1}{4}\begin{bmatrix} 10 & -4 \\ -4 & 2 \end{bmatrix} \tag{3.9}$$

The constants b_σ and θ^2, (see (2.5) and (2.8)) are equal to the max x_1^2, subject to $(x_1 x_0)G(x_1 x_0)^T = 1$. The result is

$$\theta^2 = b_\sigma = 2 . \tag{3.10}$$

The eigenvalues of G are $(3 \pm \sqrt{8})/2$. Hence

$$G(Z_0) \le (\|z_0\|^2 + \|z_1\|^2)(3 + \sqrt{8})/2 .$$

The functions appearing in the variational equations (1.15) and (1.16) are,

$$L(t) = -y'''(t)/3 , \quad p_n^{**} = h^4 y^{IV}(t_n)/4 + \mathcal{O}(h^5) .$$

4. Comments, extensions and modifications

The choice of *initial error vector* Z_0 is an obscure point of this presentation. By definition,

$$z_i = y_i^* - y(t_i) - h^p e(t_i) \qquad i = 0,1,\dots,k-1,$$

where y_i^* are the actual initial values in the computation, $y(t_i)$ are values of the correct smooth solution of (1.1), and $e(t_i)$ are values of a solution of the variational equation (1.15), which should be smooth in order that $\|p_n^{**}\| \ll \|h^{p+1}L(t)\|$. Unfortunately, we cannot for example put $z_0 = 0$ even if $y_0^* = y(t_0)$, because $e(t_0) = 0$ is usually not a starting point for a smooth solution of (1.15). According to the theory in [3] we can in that case expect $e(t_0)$ to be of the order of magnitude of $\|y'(t_0)\| \alpha |\lambda|^{-2}$ or $\|y'(t_0)\| \alpha |\lambda|^{-1}$, depending on the form of leading term of the local truncation error. The former alternative holds for the *BDF* method, while the latter holds for the implicit midpoint method. Here $|\lambda|$ is the smallest among the "large" eigenvalues of the initial Jacobian $f'(y(t_0))$. (We call an eigenvalue "large" if $|h\lambda| \geq 1$ (say).) In the autonomous case, α is a measure of the coupling from the eigenvectors belonging to the small eigenvalues to those of the large eigenvalues, due to the variation with t of the Jacobian $f'(y(t))$. When there is a gap in the eigenvalues, so that either $|h\lambda_i| \ll 1$ or $|h\lambda_j| \gg 1$ or when the coupling is weak then this is not so much a problem.

Another objection is that *the theory does not give justice to the exponential decay of the stiff components*, when L-stable methods are applied.

This can be helped if G can be chosen such that $\alpha^2 G - B^T G B$ is positive definite for some $\alpha < 1$, see [5]. Unfortunately, because of (3.8) and the rank statements of Theorem 3.2 this "strong G-stability" is possible only when $p = 1$.

A further comment is that *the theory cannot be used for methods which are not A-stable*. A modification of the error analysis is actually possible, where (2.2) is replaced by the inequality,

$$X_1^T G X_1 - X_0^T G X_0 \leq \rho(E)x_0 \cdot \sigma(E)x_0 + \mu_1\left(\sigma(E)x_0\right)^2 \tag{4.1}$$

with some reasonably small positive μ_1. The useful restrictions in the search for G given in Theorem 3.2 are then not valid, and we have not yet tried a systematic method to construct an optimal G. For the third order *BDF* method which is not far from A-stable, I found by some ad hoc sub-optimizations a matrix G, for $\mu_1 = 4/9$. This does not seem very useful, but it can probably be improved.

I believe that *a different approach* is better for such extensions, where the system is partitioned into a "stiff" and a "non-stiff" subsystem,

$$\begin{aligned} D\,dx/dt &= f(x,y) \\ dy/dt &= g(x,y) \end{aligned} \tag{4.2}$$

Here D is a small diagonal matrix, which may be singular. The upper subsystem is then to be characterized by an upper bound for $\|(f'_x)^{-1}D\|$.

The partitioning can be performed either in the analysis or in the actual computation. In the former case the same numerical method is to be used for both parts. In the latter case, a non-stiff method, for example an explicit method, can be used for the lower system.

Such a partitioning is, of course, an old idea. It is often used in connection with the *PSA* method (i.e. the pseudo-stationary approximation where one puts $D=0$). I would like to recommend that this is used, as often as possible also when the implicit methods for stiff equations are applied - in particular when the stiff part of the system is relatively small. One saves computing time and memory reducing the size of the Jacobian. The partitioning is not as critical as for the *PSA* method - because the implicit method works, also if some non-stiff components happen to be included in the upper part.

In a package designed for handling algebraic-differential systems, a low order *BDF* method can be looked upon as a backward difference correction to the *PSA* method, which (as a rule) does not complicate the system very much.

These remarks are inspired by a recent experience with some problems of chemical kinetics, where it was found that a systematic approach to the scaling, which was highly desirable for the choice of an adequate norm in the stepsize-controlling measurement of the local error (a rather neglected subject!), yielded as a by-product an adequate partitioning of the system.

This type of error analysis has not yet been fully developed, and we shall only consider a first step [5]. Suppose that the method is A_∞-stable [9], and that α, $0<\alpha<1$, is an upper bound of the moduli of the zeros of $\sigma(\zeta)$. By Stein's theorem, applied to the companion matrix of $\sigma(\zeta)/\beta_k$ there exists a positive definite matrix H, such that

$$\sigma(E)x_0 = 0 \Rightarrow X_1^T H X_1 \le \alpha^2 X_0^T H X_0 . \tag{4.3}$$

By the theory of pairs of quadratic forms, [7] this implies that there exists a positive number λ, such that the quadratic form

$$|\sigma(E)x_0|^2 - \lambda\left(X_1^T H X_1 - \alpha^2 X_0^T H X_0\right)$$

is non-negative definite. By Lemma 2.2 this implies that for all vectors $z_{n+k},\dots,z_n$, if H is defined similarly to G (see Theorem 2.1),

$$\|\sigma(E)z_n\|^2 \ge \lambda\left(\mathrm{H}(Z_{n+1}) - \alpha^2\,\mathrm{H}(Z_n)\right) \tag{4.4}$$

Now consider (1.17) and put

$$f\left(\sigma(E)y_n^*\right) - f\left(\sigma(E)y_n^{**}\right) = A_n\sigma(E)(y_n^* - y_n^{**}) = A_n\sigma(E)z_n ,$$

and suppose that

$$\|A_n^{-1}\| \le \gamma^{-1} . \tag{4.5}$$

Then, by (1.17),

$$\sigma(E)z_n = (hA_n)^{-1}\,(\rho(E)z_n - q_n)\,.$$

Hence

$$\lambda\Big(H(Z_{n+1}) - \alpha^2 H(Z_n)\Big) \le (h\gamma)^{-2}\cdot 2\Big(\|\rho(E)z_n\|^2 + \|q_n\|^2\Big)\,.$$

It is easy to determine constants b_1, b_0, such that

$$\|\rho(E)z_n\|^2 \le b_1 H(Z_{n+1}) + b_0 H(Z_n)\,.$$

Put

$$\delta^2 := 2(h\gamma)^{-2}/\lambda\,.$$

It follows that

$$(1 - b_1\delta^2)\,H(Z_{n+1}) \le (\alpha^2 + b_0\delta^2)\,H(Z_n) + \delta^2\,\|q_n\|^2\,.$$

If

$$(b_1 + b_0)\delta^2 < 1 - \alpha^2\,, \quad \|q_n\| \le \|q\| \quad (n = 0,1,2,\ldots) \tag{4.6}$$

then

$$H(Z_{n+1}) \le c^2 + \left(\frac{\alpha^2 + b_0\delta^2}{1 - b_1\delta^2}\right)^{n+1} \max\Big(0, H(Z_0) - c^2\Big) \tag{4.7}$$

where

$$c = \|q\|\,\delta\cdot\Big(1 - \alpha^2 - (b_1 + b_0)\delta^2\Big)^{-1/2}$$

This error bound, which requires only A_∞-stability together with the conditions (1.5), (4.5) and (4.6), can be applied to the "stiff" part of the system, while the previous analysis, or something else, is applied to the other part.

REFERENCES

1. Coppel, W.A., *Stability and Asymptotic Behavior of Differential Equations.* Boston, Mass.: D.C. Heath, (1965).
2. Dahlquist, G., "A Special Stability Problem for Linear Multistep Methods", *BIT*,3,27-43, (1963).
3. Dahlquist, G., "The Sets of Smooth Solutions of Differential and Difference Equations", *Stiff Differential Systems* (ed. R. Willoughby), Plenum Press, (1974).
4. Dahlquist, G., *On Stability and Error Analysis for Stiff Non-Linear Problems*, Part I, Dept. of Comp. Sci. Roy. Inst. of Technology, Report TRITA-NA-7508, (1975).
5. Dahlquist, G., *On Stability and Error Analysis for Stiff Non-Linear Problems*, Part II, Report TRITA-NA-7509, (1975).
6. Desoer, C.A. and H. Haneda, "The Measure of a Matrix as a Tool to Analyze Computer Algorithms for Circuit Analysis", *IEEE Trans. CT*-19, 480-486, (1972).
7. Hestenes, M.R., "Pairs of Quadratic Forms", *Lin. Alg. and its Appl.*, 397-407, (1968).
8. Karasalo, I., *On Smooth Solutions to Stiff Nonlinear Analytic Differential Systems*, Dept. of Comp. Sci., Roy. Inst. of Technology, Report TRITA-NA-7501, (1975).
9. Liniger, W., "Connections Between Accuracy and Stability Properties of Linear Multistep Formulas", *CACM*,18, 53-56, (1975).
10. Ortega, J.M. and W.C. Rheinboldt, *Iterative Solution of Nonlinear Equations in Several Variables*, New York: Acad. Press, (1970).

Conjugate Gradient Methods for Indefinite Systems

R Fletcher

1. Introduction

Conjugate gradient methods have often been used to solve a wide variety of numerical problems, including linear and nonlinear algebraic equations, eigenvalue problems and minimization problems. These applications have been similar in that they involve large numbers of variables or dimensions. In these circumstances any method of solution which involves storing a full matrix of this large order, becomes inapplicable. Thus recourse to the conjugate gradient method may be the only alternative. For problems in linear equations, the conjugate gradient method requires that the coefficient matrix A, is not only symmetric but also positive definite. This restriction is also implicit in applications to unconstrained minimization. Yet there are many problems in which non-trivial equations have to be added to what would otherwise be an elliptic system. One example is the restriction to divergence-free vectors in fluid dynamics (linear or nonlinear), and there are various other examples from partial differential equations. Also minimization problems in general, when involving linear or nonlinear equality constraints, provide further examples. In all these cases the coefficient matrix of the linear model has the form

$$A = \begin{bmatrix} G & -C \\ -C^T & 0 \end{bmatrix} . \qquad (1.1)$$

It is usually possible to augment the quadratic objective function (see (2.5) below) with a penalty term, so that the submatrix G is positive definite. However the matrix A is clearly indefinite (that is neither positive nor negative semi-definite) except in trivial cases.

It is usually possible to apply the conjugate gradient algorithm to indefinite systems, the difficulty being the possibility of division by zero, and hence breakdown of the algorithm. There is also the possibility of substantial error growth when this situation is almost achieved. This paper will review some suggestions which have been made for modifying the conjugate gradient method, or proposing similar types of method, to deal with symmetric indefinite problems. Attention is restricted to deriving methods which do not break down, and which can be implemented using only a few vectors of storage. One possible approach is the hyperbolic pairs method of Luenberger [5], which is shown to be unsatisfactory as it stands. A new formulation is proposed which reduces to the hyperbolic pairs method in the limiting case; nonetheless the best way of applying the method is not entirely clear, and there are reasons to think that the stability problem is not completely solved.

Another type of method is that in which the sum of squares of residuals is minimized. Two ways which have been proposed for doing this are described, and it is shown that one is to be preferred. It provides a viable algorithm for application to indefinite problems. However the algorithm is not entirely convenient for application to certain types of problem. The algorithm is also shown to be directly related to others which have been proposed, in particular the method of biconjugate gradients. This latter method is described in some detail.

It is also shown that another different algorithm can be derived from the biconjugate gradient algorithm, and which is suitable for indefinite systems. Relationships are stated which exist between these methods derived from biconjugate gradients, and the standard conjugate gradient method. In addition some recent work by Paige and Saunders [6] is reviewed, in which a new method is generated, based on the explicit use of the orthogonal reduction of a tridiagonal matrix to upper triangular form. The formulae which are entailed in using this method are quite complicated. However it is shown that Paige and Saunders' method is equivalent to the second method based on the biconjugate gradient algorithm, although the latter provides the more convenient recurrence relations. Finally a few other possibilities which have been suggested, and which are potentially suitable for the indefinite problem are discussed. None of them are currently felt to be as promising as the two methods derived from biconjugate gradients, for this type of application.

2. The Conjugate Gradient Method

The conjugate gradient method can be applied to solving the system $Ax = b$, where $x, b \in R^n$ and where A is symmetric. The further assumption that A is positive definite guarantees that the algorithm is well defined, apart from (important) considerations of round-off error. The method involves recurrence relationships, so it is convenient to use $x_1, x_2, \dots$ to denote successive iterates. It is also convenient to define a residual $r(x) \equiv b - Ax$, and r_i is often used to refer to $r(x_i)$, although this usage is corrupted at certain times. The algorithm is described in detail by Hestenes and Stiefel [2], and can be stated in various equivalent forms. All require x_1 and hence $r_1 \equiv r(x_1)$ to be given, and set $p_1 = r_1$. Then for $k = 1,2,\dots$, recurrence relations are used, one form of which is

$$x_{k+1} = x_k + \alpha_k p_k \qquad (2.1a)$$

$$r_{k+1} = r_k - \alpha_k A p_k \qquad (2.1b)$$

where

$$\alpha_k = r_k^T r_k / p_k^T A p_k \qquad (2.1c)$$

and

$$p_{k+1} = r_{k+1} + \beta_k p_k \qquad (2.1d)$$

where

$$\beta_k = r_{k+1}^T r_{k+1} / r_k^T r_k \, . \qquad (2.1e)$$

The algorithm terminates in exact arithmetic in at most n iterations (m say) with $r_{m+1} = 0$ and $\beta_m = 0$. Four storage vectors are required to implement the method on a computer. The properties of the method will be stated here without explicit proof; (proof is given in [2] and also in section 5 as a special case of the proof of the biconjugate gradient algorithm.) If the Krylov sequence is defined as

$$\{r_1, Ar_1, A^2 r_1, \ldots, A^{k-1} r_1\} \, , \qquad (2.2)$$

then for $k \leq m$ this is a basis for a linear space S_k. Then r_k and p_k are in S_k but not S_{k-1}, the vector r_k satisfying orthogonality conditions

$$r_k^T r_j = r_k^T p_j = 0 \qquad \text{for all} \quad j < k \, , \qquad (2.3)$$

and the vector p_k satisfying conjugacy conditions

$$p_k^T A p_j = 0 \qquad \text{for all} \quad j < k \, . \qquad (2.4)$$

The vectors $r_1, r_2, \ldots, r_k$ are therefore an orthogonal basis for S_k.

An alternative interpretation is that $-r(x)$ is the gradient of the quadratic function

$$Q(x) = \tfrac{1}{2} x^T A x - b^T x \qquad (2.5)$$

and p_{k+1} is the component of r_{k+1} which is conjugate to p_k - hence the term conjugate gradients. Also the choice of α_k is that which makes $Q(x_k + \alpha p_k)$ stationary with respect to distance α along a direction of search p_k. Thus it is possible to rearrange equations (2.1) so that Ap_k is not evaluated explicitly, but r_{k+1} is calculated as $r(x_{k+1})$ and α_k is obtained by solving the equation $r(x_{k+1})^T p_k = 0$. In this interpretation the extension to general minimization problems becomes apparent. Of course for nonlinear problems the termination property no longer holds. However this result is largely illusory since the method suffers from build up of round-off error when k becomes large. Also in application to very large systems, even carrying out n iterations can be prohibitive. However, regarding the method as an iterative method (for linear or nonlinear systems), it has been observed that for certain systems, satisfactory convergence occurs in much fewer than n iterations. There are good reasons for this when the system satisfies certain symmetry properties (see Reid [7], for example), and it is under these circumstances that conjugate gradients is viable. However it is largely a matter of trial and error to find out whether any given system can be solved rapidly.

Further useful properties relating to conjugate gradients are obtained by defining the matrices $(\|.\| \equiv \|.\|_2)$

$$\mathcal{R}_k = \left[\frac{r_1}{\|r_1\|}, \frac{r_2}{\|r_2\|}, \ldots, \frac{r_k}{\|r_k\|}\right] \tag{2.6}$$

$$\mathcal{P}_k = \left[\frac{p_1}{\|r_1\|}, \frac{p_2}{\|r_2\|}, \ldots, \frac{p_k}{\|r_k\|}\right] \tag{2.7}$$

$$\mathcal{L}_k = \begin{bmatrix} 1 & & & & \\ -\beta_1^{\frac{1}{2}} & 1 & & & \\ & -\beta_2^{\frac{1}{2}} & 1 & & \\ & & \ddots & \ddots & \\ & & & -\beta_{k-1}^{\frac{1}{2}} & 1 \end{bmatrix} \tag{2.8}$$

and

$$\mathcal{D}_k = \begin{bmatrix} \alpha_1^{-1} & & & \\ & \alpha_2^{-1} & & \\ & & \ddots & \\ & & & \alpha_k^{-1} \end{bmatrix} \tag{2.9}$$

Note that $\mathcal{R}_n$ is an orthogonal matrix if $m = n$. Using the fact that $r_{m+1} = 0$, the recurrence relation (2.1b) can be written

$$A\mathcal{P}_m\mathcal{D}_m^{-1} = \mathcal{R}_m\mathcal{L}_m , \tag{2.10}$$

and using $p_1 = r_1$, (2.1d) can be written

$$\mathcal{P}_m\mathcal{L}_m^T = \mathcal{R}_m . \tag{2.11}$$

Eliminating $\mathcal{P}_m$ from (2.10), and defining $T_k = \mathcal{L}_k\mathcal{D}_k\mathcal{L}_k^T$, gives the equation

$$A\mathcal{R}_m = \mathcal{R}_m\mathcal{L}_m\mathcal{D}_m\mathcal{L}_m^T = \mathcal{R}_m T_m \tag{2.12}$$

showing that when $m = n$, $\mathcal{R}_n$ provides an orthogonal transformation of A to symmetric tridiagonal form T_n. The equations implicit in (2.12) were used by Lanczos [3] as the basis of a method for finding eigenvalues of A. Any normalized vector r_1 is chosen (not $r(x_1)$ as when solving equations) and the three term recurrence implicit in (2.12) (rearranged to give r_{k+1} on the left) is carried out for all k (assuming $m = n$). Then the elements of T_n and hence the eigenvalues of A can be determined. Note that any subsequent determination of eigenvectors by inverse iteration might require the solution of a symmetric indefinite system of linear equations. In what follows, it is also convenient to have an expression for $A\mathcal{R}_k$ for an intermediate stage $k < m$. Let the extension $\tilde{\mathcal{L}}_k$ of $\mathcal{L}_k$ be defined as the first k columns of $\mathcal{L}_{k+1}$ (so that $(\tilde{\mathcal{L}}_k; e_{k+1}) \equiv \mathcal{L}_{k+1}$ where e_i denotes the i-th coordinate vector). Then (2.1b) becomes

$$A\mathcal{P}_k\mathcal{D}_k^{-1} = \mathcal{R}_{k+1}\tilde{\mathcal{L}}_k \tag{2.13}$$

and using (2.11), then

$$A\mathcal{R}_k = \mathcal{R}_{k+1}\tilde{\mathcal{L}}_k\mathcal{D}_k\mathcal{L}_k^T = \mathcal{R}_{k+1}\tilde{T}_k \tag{2.14}$$

say, where $\tilde{T}_k$ is the <u>extension</u> of T_k. The result that

$$x_{k+1} - x_1 = \sum_{i=1}^{k} \alpha_i p_i$$

$$= \mathcal{R}_k T_k^{-1} e_1 \|r_1\| \tag{2.15}$$

can be derived readily, using (2.11) and the definition of T_k.

In applying conjugate gradients to elliptic systems, A is positive definite, so all the matrices T_k are positive definite, and there is no difficulty with the nonexistence of the factors $\mathcal{L}_k$ and $\mathcal{D}_k$, or of the iterates x_k. However when A is indefinite, then although T_k exists for all k, T_k^{-1} and hence x_{k+1} may not exist or may be large, and also the factors may not exist for certain k, even when T_k is non-singular. The search for new methods may therefore be regarded as looking for different simple factorizations of either A or T_n.

3. Hyperbolic pairs

The precise way in which the conjugate gradient algorithm can fail for an indefinite system is that in calculating α_k by (2.1c) the denominator satisfies

$$p_k^T A p_k = 0 . \tag{3.1}$$

Luenberger [5] has shown that the algorithm can be rescued when this occurs, by exploiting the concept of a <u>hyperbolic pair</u>. He defines a vector p_{k+1} as a linear combination of Ap_k and p_k such that

$$p_{k+1}^T A p_{k+1} = 0 \tag{3.2}$$

and (3.1) and (3.2) constitute the definition of p_k and p_{k+1} as a hyperbolic pair. Then scalars α_k and α_{k+1} can be calculated by formulae different from (2.1c) giving $x_{k+2} = x_k + \alpha_k p_k + \alpha_{k+1} p_{k+1}$ (see (3.4) below), with an analogous formula for r_{k+2}. A formula is also given for p_{k+2} which therefore enables the algorithm to be continued using the recurrence relations (2.1).

Luenberger's algorithm is unsuitable in general because it does not adequately cope with the situation when $p_k^T A p_k$ is small (ε say) but not zero. Then x_{k+1} is still calculated by (2.1a) and can be arbitrarily large. This will introduce relatively large rounding errors into the subsequent calculation.

It is also bad in a nonlinear situation, since the information contained in r_{k+1} is worthless. However it will now be shown that further modifications can be introduced into the algorithm to cater for this situation to some extent. It is easy to verify that when $p_k^T A p_k$ tends to zero, then p_{k+1} tends to p_k in direction. In fact although the point x_{k+1} tends to infinity, the point x_{k+2} is well-defined in the limit. It is therefore possible to rearrange the computation so that x_{k+2} is calculated directly from x_k, and so that the need to compute x_{k+1}, r_{k+1} and p_{k+1} is removed. To do this it is convenient to write

$$\bar{r}_{k+1} = \alpha_k^{-1} r_{k+1} = \alpha_k^{-1} r_k - Ap_k \,. \tag{3.3a}$$

Then it can be verified that

$$x_{k+2} - x_k = -\frac{r_k^T r_k}{\bar{r}_{k+1}^T \bar{r}_{k+1}} \left\{ \frac{\bar{r}_{k+1}^T A \bar{r}_{k+1}}{\bar{r}_{k+1}^T \bar{r}_{k+1}} p_k + \bar{r}_{k+1} \right\} / (1-\mu_k) \tag{3.3b}$$

where

$$\mu_k = \alpha_k^{-1} \left(\frac{r_k^T r_k}{\bar{r}_{k+1}^T \bar{r}_{k+1}} \right) \left(\frac{\bar{r}_{k+1}^T A \bar{r}_{k+1}}{\bar{r}_{k+1}^T \bar{r}_{k+1}} \right) . \tag{3.3c}$$

In addition it also can be shown that

$$p_{k+2} = r_{k+2} + \frac{r_{k+2}^T r_{k+2}}{r_k^T r_k} \left\{ \frac{p_k^T A p_k}{\bar{r}_{k+1}^T \bar{r}_{k+1}} \bar{r}_{k+1} + p_k \right\} . \tag{3.3d}$$

Thus if it is decided that $p_k^T A p_k$ is too small to permit the recurrences in (2.1) to be used in a stable fashion, the alternative is to calculate $\bar{r}_{k+1}$ by (3.3a). Then $A\bar{r}_{k+1}$ must be calculated, in which case μ_k, x_{k+2}, $r_{k+2} \equiv r(x_{k+2})$ and p_{k+2} follow. To use this algorithm in a nonlinear situation without A being explicitly available, requires an additional search along $\bar{r}_{k+1}$ to calculate $A\bar{r}_{k+1}$. When $p_k^T A p_k = 0$, then $\alpha_k^{-1} = 0$ and hence $\bar{r}_{k+1} = -Ap_k$ and $\mu_k = 0$. Then (3.3b) and (3.3d) reduce to

$$x_{k+2} - x_k = \frac{r_k^T r_k}{p_k^T A^2 p_k} \left\{ Ap_k - \frac{p_k^T A^3 p_k}{p_k^T A^2 p_k} p_k \right\} \tag{3.4}$$

and

$$p_{k+2} = r_{k+2} + \frac{r_{k+2}^T r_{k+2}}{r_k^T r_k} p_k \,. \tag{3.5}$$

Although these formulae are more simple than those given by Luenberger, it is not difficult to verify that they are equivalent.

One difficult with using this algorithm is that of deciding when $p_k^T A p_k$ is sufficiently small to merit using the formulae (3.3), rather than (2.1). However the algorithm has apologies with the algorithm of Bunch and Parlett [1] for factorizing a symmetric matrix efficiently using either 1×1 or 2×2

symmetric block diagonal pivots (corresponding to the one step (2.1) or two step (3.3) formulae here). In this case the symmetric matrix in question is T_n and the algorithm has to be applied without interchanges. With this restriction in mind, it might be possible to adapt the test of Bunch and Parlett to work here. Unfortunately using 1×1 or 2×2 pivots to factorize a tridiagonal matrix without interchanges is still potentially unstable to some extent. Because of this the methods developed in sections 4 and 5 seem preferable.

4. Least squares and minimum residuals

Another possible approach towards determining a stable algorithm is to attempt to minimize the sum of squares of residuals, that is $r(x)^T r(x)$. In the linear case the necessary and sufficient conditions for this are that the normal equations of least squares, $A^2x = Ab$, are satisfied, when A is symmetric and non-singular. It is possible to apply the algorithm (2.1) to this system, since A^2 is positive definite. However there are two disadvantages to this approach. One is that two multiplications by A are required per iteration, which roughly doubles the amount of work, since there is no reason to expect fewer iterations to be required. The other reason is that severe ill-conditioning often becomes apparent. This is related to the fact that the Krylov sequence for the least squares system is $\{Ar_1, A^3r_1, A^5r_1, \ldots\}$, which is much harder to resolve numerically into orthogonal vectors than the sequence $\{r_1, Ar_1, A^2r_1, \ldots\}$ for conjugate gradients applied to $Ax = b$, by analogy with the power method for computing eigenvalues of A.

However Hestenes and Stiefel [2] point out that there exist vectors $\{x_k\}$ say, in the convex hull of the iterates, $\{x_k^{CG}\}$ generated by the conjugate gradient method (2.1), whose residuals r_k minimize r^Tr for r in the space $\{r_1^{CG}, Ar_1^{CG}, A^2r_1^{CG}, \ldots, A^{k-1}r_1^{CG}\}$, and which satisfy a simple recurrence. Choosing this sequence of vectors $\{x_k\}$ provides an algorithm which does not suffer from the disadvantages above, and which is termed the minimum residual algorithm. The same algorithm can be derived in different ways by using the generalization of conjugate gradients given by Rutishauser [8] in which the general inner product (p,q) is defined as $p^TA^{\mu}q$, for any integer μ. In this case $\mu = 1$ gives the minimum residual algorithm. This is easy to see since the conditions $(r_k^Tp_j) = 0$ for $j < k$, become $r_k^TAp_j = 0$ when the innerproduct is redefined with $\mu = 1$. But the gradient $\nabla(r^Tr) = -2Ar$ by definition of $r(x)$, so it follows that $\nabla(r^Tr)|_{x_k}$ is orthogonal to p_j for all $j < k$, and so r^Tr is minimized in the space spanned by these $\{p_j\}$. Since only the innerproduct is redefined, it is easy to verify that this space is the same as in standard conjugate gradients.

For indefinite systems the minimum residual algorithm is not stable, in the form given on redefining the innerproduct and using (2.1). This is because $(r_k, r_k) \equiv r_k^T A r_k$ can become zero in the definition of β_k. However this difficulty can be overcome by rearranging the algorithm suitably as will be demonstrated in section 5. In this rearranged form, the algorithm requires 6 vectors of storage for a computer program. One disappointing feature of the minimum residual algorithm is that it is not entirely convenient when A is not available (such as in applications to minimization subject to constraints). This is because in determining $p_{k+1} = r_{k+1} + \beta_k p_k$, the computation of $\beta_k = r_{k+1}^T A r_{k+1} / r_k^T A r_k$ (or its equivalent when the algorithm is rearranged) requires Ar_{k+1} which is not available, so has to be obtained by searching along r_{k+1}. Thus each iteration $(k > 1)$ requires a search first along r_{k+1} (or alternatively Ap_k), before the search along p_{k+1} can be made.

5. Biconjugate Gradients

In the search for a suitable algorithm for solving indefinite systems it is fruitful to examine a different generalization of conjugate gradients. This is the biconjugate gradient algorithm, used by Lanczos [3] as a means of computing the eigenvalues of an unsymmetric matrix A. In this form, two vectors r_1 and $\bar{r}_1$ are given, and letting $p_1 = r_1$ and $\bar{p}_1 = \bar{r}_1$, then for $k = 1,2,\ldots$, the recurrence relations

$$r_{k+1} = r_k - \alpha_k A p_k \tag{5.1a}$$

$$\bar{r}_{k+1} = \bar{r}_k - \alpha_k A^T \bar{p}_k \tag{5.1b}$$

where

$$\alpha_k = \bar{r}_k^T r_k / \bar{p}_k^T A p_k , \tag{5.1c}$$

and

$$p_{k+1} = r_{k+1} + \beta_k p_k \tag{5.1d}$$

$$\bar{p}_{k+1} = \bar{r}_{k+1} + \beta_k \bar{p}_k \tag{5.1e}$$

where

$$\beta_k = \bar{r}_{k+1}^T r_{k+1} / \bar{r}_k^T r_k , \tag{5.1f}$$

are used. The scalar α_k is chosen so as to force the biorthogonality conditions $r_{k+1}^T \bar{r}_k = \bar{r}_{k+1}^T r_k = 0$ and β_k is chosen to force the biconjugacy condition $\bar{p}_{k+1}^T A p_k = p_{k+1}^T A^T \bar{p}_k = 0$. It is a feature of the algorithm (analogous to the algorithm (2.1)) that these conditions also hold for any pair of vectors without having to be explicitly enforced. This can be stated formally as follows.

Theorem Assuming that the biconjugate gradient algorithm (5.1) does not break down, then for all $j < i$,

$$\bar{r}_i^T r_j = r_i^T \bar{r}_j = 0 \tag{5.2}$$

and

$$\bar{p}_i^T A p_j = p_i^T A^T \bar{p}_j = 0 . \tag{5.3}$$

Proof Clearly the theorem holds for $i = 1$ since no condition is implied. For $i > 1$ an inductive argument is used, assuming that the theorem is true for i, and proving it when i is replaced by $i+1$. Firstly by (5.1a)

$$r_{i+1}^T \bar{r}_j = r_i^T \bar{r}_j - \alpha_i p_i^T A^T \bar{r}_j = r_i^T \bar{r}_j - \alpha_i p_i^T A^T (\bar{p}_j - \beta_{j-1} \bar{p}_{j-1}) \qquad (5.4)$$

by (5.1e), assuming $\beta_0 = 0$ if necessary. For $j=i$, it follows from (5.3) and (5.1c) that $r_{i+1}^T \bar{r}_i = 0$. For $j < i$ in (5.4), it follows from (5.2) and (5.3) that $r_{i+1}^T \bar{r}_j = 0$, so this result holds for all $j < i+1$. In a like manner $\bar{r}_{i+1}^T r_j = 0$ can be demonstrated so it follows that

$$\bar{r}_{i+1}^T r_j = r_{i+1}^T \bar{r}_j = 0 \qquad \text{for all} \quad j < i+1, \qquad (5.5)$$

which is (5.2) with i replaced by $i+1$.

It also follows from (5.1d) and (5.1b) that

$$p_{i+1}^T A^T \bar{p}_j = r_{i+1}^T A^T \bar{p}_j + \beta_i p_i^T A^T \bar{p}_j$$

$$= r_{i+1}^T (\bar{r}_j - \bar{r}_{j+1})/\alpha_j + \beta_i p_i^T A^T \bar{p}_j \,. \qquad (5.6)$$

When $j = i$, using (5.1c) gives

$$p_{i+1}^T A^T \bar{p}_i = r_{i+1}^T (\bar{r}_i - \bar{r}_{i+1})/\alpha_i + \beta_i \, r_i^T \bar{r}_i / \alpha_i \,.$$

Then by (5.5) and (5.1f), $p_{i+1}^T A^T \bar{p}_i = 0$. When $j < i$ in (5.6), then using (5.5) and (5.3) gives $p_{i+1}^T A^T \bar{p}_j = 0$, so this result holds for all $j < i+1$. Likewise $\bar{p}_{i+1}^T A p_j = 0$ can be established for all $j < i+1$, and these results extend (5.3) when i is replaced by $i+1$.

Q.E.D.

A direct consequence of (5.1d) is that $p_{k+1} = r_{k+1} + \beta_k r_k + \beta_k \beta_{k-1} r_{k-1} + \ldots$, so that (5.2) also implies that

$$r_i^T \bar{p}_j = \bar{r}_i^T p_j = 0 \qquad (5.7)$$

for $j < i$. Another simple corollary is that the algorithm must terminate with $r_{m+1} = \bar{r}_{m+1} = 0$ in at most n iterations, using the same argument as for the standard conjugate gradient method. It may seem unusual to state these results in detail for the biconjugate gradient algorithm but not for other conjugate gradient methods. However as will be demonstrated, almost all algorithms which are considered here are special cases of this algorithm, so these results carry over immediately.

In the form (5.1) the algorithm provides a similarity reduction of A to tridiagonal form, suitable for calculating eigenvalues. To solve a system of equation $Ax = b$, it is necessary to augment (5.1). To do this r_1 is chosen to be the residual $r(x_1)$, where x_1 is an initial estimate of the solution. Then the recurrence relation

$$x_{k+1} = x_k + \alpha_k p_k \qquad (5.8)$$

is carried out together with (5.1), and since $r_{m+1} = 0$, it follows that x_{m+1} solves the equations. For general unsymmetric A there is no guarantee that the algorithm will not break down or not be unstable.

In the applications of interest here, A is a symmetric matrix. In this case the choice of $\bar{r}_1 = r_1$ leads to the standard conjugate gradient algorithm (2.1), and $\bar{r}_k = r_k$, $\bar{p}_k = p_k$ for all k. However in this section the main interest will lie in examining the consequences of making the choice

$$\bar{r}_1 = Ar_1 . \tag{5.9}$$

It then follows easily when A is symmetric that

$$\bar{r}_k = Ar_k , \qquad \bar{p}_k = Ap_k \tag{5.10}$$

for all k. Since all innerproducts in (5.1) are between either $(\bar{r}_k, r_k)$ or $(\bar{p}_k, Ap_k)$, it follows from (5.10) that the resulting algorithm is equivalent to the standard conjugate gradient algorithm but with (a,b) defined as $a^T Ab$. Thus the biconjugate gradient algorithm (5.1) and (5.8), with the choice (5.9), is equivalent to the minimum residual algorithm described in section 4. That is to say the vectors x_k, r_k and p_k in (5.8), (5.1a) and (5.1d) recur as in the minimum residual algorithm. It is also possible to show (see section 6) that the vectors r_k are proportional to the search directions (p_k^{CG} say) which arise in the standard conjugate gradient algorithm (2.1).

Before pursuing these ideas further, it is noted that the choices $\bar{r}_1 = r_1$ and $\bar{r}_1 = Ar_1$ are special cases of a more general class of methods. Consider choosing $\bar{r}_1 = Mr_1$ where M is any square matrix. Then a sufficient condition for the properties $\bar{r}_k = Mr_k$, $\bar{p}_k = Mp_k$ to persist is that

$$A^T M = MA \tag{5.11}$$

This is easily proved by induction. When A is symmetric, (5.11) is the commutation condition $AM = MA$. The resulting method is then equivalent to using the generalized innerproduct $(a,b) \equiv a^T Mb$ in (2.1). An example of a matrix which commutes is $M = A^{\mu}$, which has already been referred to, but it can be seen for instance that any matrix polynomial $M = \Pi(A)$ also commutes. A choice of possible interest might be the general linear polynomial $M = \lambda A + \mu I$.

Attention will henceforth be restricted to the case when A is symmetric and the choice $\bar{r}_1 = Ar_1$ is made. It will be noticed that the minimum residual algorithm makes no reference to the vectors $\{\bar{p}_k\}$ which are generated in the biconjugate gradient algorithm. Furthermore these vectors are mutually <u>orthogonal</u>, since the conditions (5.10) and the biconjugacy conditions (5.3) imply that $\bar{p}_i^T \bar{p}_j = \bar{p}_i^T Ap_j = 0$, when $j \neq i$. It is interesting therefore to consider whether any further algorithms can be developed, in which the vectors $\bar{p}_k$ are used as search directions, especially since stability might be expected to arise from the orthogonality property. Consider then introducing a new sequence of iterates $\{\hat{x}_k\}$ with $\hat{x}_1 = x_1$, obtained by searching the directions $\bar{p}_k$, that is by

$$\hat{x}_{k+1} = \hat{x}_k + \hat{\alpha}_k \bar{p}_k \tag{5.12a}$$

where $\hat{\alpha}_k$ is yet to be determined. Then the residuals $\hat{r}_k \equiv r(\hat{x}_k)$ recur like

$$\hat{r}_{k+1} = \hat{r}_k - \hat{\alpha}_k A\bar{p}_k \,. \tag{5.12b}$$

There is no difficulty in carrying out these recurrences in conjunction with those given by (5.1). Furthermore it is possible to choose the $\{\hat{\alpha}_k\}$ so that the sequence $\{\hat{x}_k\}$ terminates at the solution and $\{\hat{r}_k\}$ at zero. In fact $\hat{\alpha}_k$ is chosen so that $\hat{r}_{k+1}^T r_k = 0$, giving the equation

$$\hat{\alpha}_k = \hat{r}_k^T r_k / \bar{p}_k^T A r_k \tag{5.13}$$

or equivalently, since $r_k = p_k - \beta_{k-1} p_{k-1}$ and using (5.3), by

$$\hat{\alpha}_k = \hat{r}_k^T r_k / \bar{p}_k^T \bar{p}_k \,. \tag{5.14}$$

A simple inductive proof shows that $\hat{r}_i^T r_j = 0$ for all $j < i$. First of all, $\hat{r}_{i+1}^T r_i = 0$ by choice of $\hat{\alpha}_i$. But for all $j < i$, $\hat{r}_{i+1}^T r_j = \hat{r}_i^T r_j - \hat{\alpha}_i \bar{p}_i^T A r_j = -\hat{\alpha}_i \bar{p}_i^T A r_j$, and eliminating $A\bar{p}_i$ using (5.1b) and using (5.2), shows that this scalar product is zero. Thus $\hat{r}_{i+1}^T r_j = 0$ for all $j < i+1$ which completes the induction. Now $\hat{r}_{n+1}^T r_j = 0$ for all $j \leq n$, implies that $\hat{r}_{n+1} = 0$, so it follows that the sequence $\{\hat{x}_k\}$ terminates at the solution in at most n iterations. Thus another algorithm has been derived from the biconjugate gradient algorithm. This new algorithm uses the recurrences (5.12a,b) and those for r_k, $\bar{r}_k$ and $\bar{p}_k$ from (5.1); note that p_k is not required.

In this form however, the algorithm fails when $\bar{r}_k^T r_k = 0$ (as also does the minimum residual algorithm). However the difficulties can be avoided by eliminating $\bar{r}_{k+1}$ from (5.1e) using (5.1b), and then eliminating $\bar{r}_k$ from the resulting expression using (5.1e). This gives $\bar{p}_{k+1}$ as a linear combination of $A\bar{p}_k$, $\bar{p}_k$ and $\bar{p}_{k-1}$, which on redefining the length of $\bar{p}_{k+1}$, can be written as

$$\bar{p}_{k+1} = A\bar{p}_k - \lambda_k \bar{p}_k - \mu_k \bar{p}_{k-1} \,. \tag{5.15}$$

By virtue of the orthogonality of the $\{\bar{p}_k\}$, the simple expressions

$$\lambda_k = \bar{p}_k^T A \bar{p}_k / \bar{p}_k^T \bar{p}_k \tag{5.16}$$

and

$$\mu_k = \bar{p}_{k-1}^T A \bar{p}_k / \bar{p}_{k-1}^T \bar{p}_{k-1} = \bar{p}_k^T \bar{p}_k / \bar{p}_{k-1}^T \bar{p}_{k-1} \tag{5.17}$$

follow readily. In this form, the algorithm requires the recurrences for $\hat{x}_k$, $\hat{r}_k$, $\bar{p}_k$ and r_k. A similar expression

$$p_{k+1} = Ap_k - \lambda_k p_k - \mu_k p_{k-1} \tag{5.18}$$

can be used to stabilize the minimum residual algorithm. However the expression (5.13) for $\hat{\alpha}_k$ is still potentially unsatisfactory. Therefore eliminating r_k using $r_k = r_{k-1} - \alpha_{k-1}\bar{p}_{k-1}$, and simplifying yields $\hat{\alpha}_k = \hat{r}_k^T\bar{p}_{k-1}/\bar{p}_k^T\bar{p}_k$ $(k>1)$. It now transpires that the recurrence r_k is no longer required in the algorithm, so an even more simple algorithm can be stated.

Given $\hat{x}_1$, set $\hat{r}_1 = r(\hat{x}_1)$, $\bar{p}_1 = A\hat{r}_1$ and let $\mu_1 = 0$ and $\bar{p}_0 = \hat{r}_1$ to allow for the initial special cases. Then repeat for $k = 1,2,\ldots$ the following

$$\hat{\alpha}_k = \hat{r}_k^T\bar{p}_{k-1}/\bar{p}_k^T\bar{p}_k \tag{5.19a}$$

$$\hat{x}_{k+1} = \hat{x}_k + \hat{\alpha}_k\bar{p}_k \tag{5.19b}$$

$$\hat{r}_{k+1} = \hat{r}_k - \hat{\alpha}_k A\bar{p}_k \tag{5.19c}$$

$$\lambda_k = \bar{p}_k^TA\bar{p}_k/\bar{p}_k^T\bar{p}_k \tag{5.19d}$$

$$\mu_k = \bar{p}_k^T\bar{p}_k/\bar{p}_{k-1}^T\bar{p}_{k-1} \tag{5.19e}$$

$$\bar{p}_{k+1} = A\bar{p}_k - \lambda_k\bar{p}_k - \mu_k\bar{p}_{k-1} \tag{5.19f}$$

This algorithm will be called the <u>orthogonal direction algorithm</u>. It requires five vectors of storage and is stable. Also the errors in the $\{\hat{x}_k\}$ decrease monotonically due to the orthogonality of the $\{\bar{p}_k\}$. The algorithm adapts well as a search algorithm for nonlinear problems, since $\hat{r}_{k+1}$ can alternatively be defined as $r(\hat{x}_{k+1})$, and then $A\bar{p}_k$ computed from differences in $\hat{r}_{k+1}$ and $\hat{r}_k$.

There are some unanswered questions about implementation, for instance about propagation of round-off error and practical conditions for termination. To this extent it might be convenient to recur the minimum residual vector r_k, at the expense of a vector of storage. Alternatively certain collinearity properties which exist between the orthogonal direction algorithm, the minimum residual algorithm and the conjugate gradient algorithm (given in section 6) might be used. When the minimum residual vector is available, there are also certain alternatives for nonlinear problems in regard to calculating $\hat{\alpha}_k$, which might be explored.

6. Interrelationships

Recently, Paige and Saunders [6] suggested another new method based on the use of a QR factorization. In this section their algorithm is described and is shown to give the same sequence of iterates $\{\hat{x}_k\}$ as the algorithm (5.19) derived from biconjugate gradients. Interrelationships between this algorithm, the minimum residual algorithm and the conjugate gradient algorithm are then stated.

Consider the analysis of section 2, and in particular the extended tridiagonal matrix $\mathbb{T}_k = \mathcal{R}_{k+1}^T A \mathcal{R}_k$ defined in the conjugate gradient method. Paige and

Saunders [6] base their method on a QR factorization of $\tilde{T}_k$. If an orthogonal matrix Q_{k+1} is defined as the product of k plane rotations in the (1,2), (2,3),...,(k,k+1) planes

$$Q_{k+1} = \begin{bmatrix} c_1 & s_1 & & & \\ s_1 & -c_1 & & & \\ & & 1 & & \\ & & & \ddots & \\ & & & & 1 \end{bmatrix} \cdots\cdots\cdots \begin{bmatrix} 1 & & & \\ & \ddots & & \\ & & 1 & & \\ & & & c_k & s_k \\ & & & s_k & -c_k \end{bmatrix}, \tag{6.1}$$

then it is possible to choose $c_1, c_2, \ldots, c_k$ to eliminate the subdiagonal elements of $\tilde{T}_k$ in order, so that

$$Q_{k+1}^T \tilde{T}_k = \begin{pmatrix} U_k \\ 0 \end{pmatrix} = \begin{bmatrix} x & x & x & & \\ & x & x & x & \\ & & x & x & x \\ & & & x & x \\ & & & & x \\ 0 & \cdots & \cdots & \cdots & 0 \end{bmatrix}. \tag{6.2}$$

Then

$$\mathcal{R}_{k+1} \tilde{T}_k = \mathcal{R}_{k+1} Q_{k+1} \begin{pmatrix} U_k \\ 0 \end{pmatrix} = W_k U_k \tag{6.3}$$

where W_k is the first k columns of $\mathcal{R}_{k+1} Q_{k+1}$. Note that the columns of W_k are mutually orthogonal. Now (2.15) shows that the iterates $\{x_k^{CG}\}$ in conjugate gradients satisfy

$$x_{k+1}^{CG} - x_1^{CG} = \mathcal{R}_k T_k^{-1} e_1 \|r_1\| . \tag{6.4}$$

If $\mathcal{R}_k T_k^{-1}$ is replaced by $\mathcal{R}_{k+1} \tilde{T}_k^{+T}$, then the solution x_{m+1}^{CG} is not affected, since $\tilde{T}_m = \begin{pmatrix} T_m \\ 0 \end{pmatrix}$ so that $\tilde{T}_m^{+T} = \begin{pmatrix} T_m^{-T} \\ 0 \end{pmatrix} = \begin{pmatrix} T_m^{-1} \\ 0 \end{pmatrix}$. However different intermediates, $\{x_k^{PS}\}$ say, are obtained, which always exist since $\tilde{T}_k$ always has full rank. It is easily verified that $\tilde{T}_k^{+T} = Q_{k+1} \begin{pmatrix} U_k^{-T} \\ 0 \end{pmatrix}$ so it follows that

$$x_{k+1}^{PS} - x_1^{PS} = W_k U_k^{-T} e_1 \|r_1\| = W_k z_k \tag{6.5}$$

say. Forward substitution in $U_k^T z_k = e_1 \|r_1\|$ gives one element of z_k (ζ_k say) at each iteration. Since W_{k-1} is a submatrix of W_k, it follows from (6.5) that

$$x_{k+1}^{PS} = x_k^{PS} + \zeta_k w_k \tag{6.6}$$

where w_k is the kth column of W_k. This is the algorithm which Paige and Saunders suggest. It requires the three-term recurrence relationship derived from (2.12) to be used, which generates the columns of $\mathcal{R}_m$ directly from A. The computation of the elements $c_1, c_2, \ldots$ in the plane rotation matrices requires square roots, and this calculation and that in forming the elements of U_k^T are quite intricate. The process apparently can be carried out using only five storage vectors (see Paige and Saunders' subroutine in [6]), but this does not explicitly include the vector $r_k^{PS} \equiv r(x_k^{PS})$. Since the algorithm does not introduce new information at each stage through Aw_k (but rather through the three-term recurrence relation for the columns of $\mathcal{R}_k$), it is not conveniently adapted as a search algorithm for nonlinear problems. However I must acknowledge a great debt to this work, since it made me re-examine the idea of searching the orthogonal directions $\{\bar{p}_k\}$ in the biconjugate gradient algorithm, which I had previously abandoned.

In demonstrating the interrelationships between the various methods, it is first pointed out that the reduction (6.2) equivalently implies that $\tilde{\mathcal{L}}_k$ (if it exists) is reduced to upper bidiagonal form on premultiplication by Q_{k+1}^T. This is most conveniently written

$$Q_{k+1}^T \tilde{\mathcal{L}}_k \mathcal{D}_k^{\frac{1}{2}} = \begin{pmatrix} B_k \\ 0 \end{pmatrix} = \begin{bmatrix} x & x & & & \\ & x & x & & \\ & & x & x & \\ & & & x & x \\ & & & & x \\ 0 & & & & 0 \end{bmatrix} . \tag{6.7}$$

Consider now the minimum residual algorithm. The vectors r_k are uniquely defined (apart from length) by being in the linear space S_k (see section 2) and satisfying the conditions $r_k^T A r_j = 0$, for all $j < k$. Therefore in the biconjugate gradient algorithm, with A symmetric and $\bar{r}_1 = Ar_1$; if the matrices

$$R_k = \left[\frac{r_1}{\rho_1}, \frac{r_2}{\rho_2}, \ldots, \frac{r_k}{\rho_k}\right] \tag{6.8a}$$

$$\bar{R}_k = \left[\frac{\bar{r}_1}{\rho_1}, \frac{\bar{r}_2}{\rho_2}, \ldots, \frac{\bar{r}_k}{\rho_k}\right] \tag{6.8b}$$

are defined, where $\rho_i = \sqrt{\bar{r}_i^T r_i}$, then R_k and $\bar{R}_k$ are uniquely defined by the conditions

$$AR_k = \bar{R}_k \tag{6.9a}$$

$$\bar{R}_k^T R_k = I \tag{6.9b}$$

$$r_k/\rho_k \in S_k \tag{6.9c}$$

for all k. It is readily verified that the definitions of R_k and $\bar{R}_k$ given by the expressions

$$\mathcal{R}_k = R_k \mathcal{D}_k^{\frac{1}{2}} \mathcal{L}_k^T \tag{6.10a}$$

and

$$\bar{R}_k = \mathcal{R}_{k+1} \tilde{\mathcal{L}}_k \mathcal{D}_k^{\frac{1}{2}} \tag{6.10b}$$

satisfy the conditions (6.9), and so relate the biconjugate gradient method to the conjugate gradient method.

Relationships can also be derived for the directions p_k and $\bar{p}_k = Ap_k$ in the biconjugate gradient algorithm. Defining

$$P_k = \left[\frac{p_1}{\|\bar{p}_1\|}, \frac{p_2}{\|\bar{p}_2\|}, \ldots, \frac{p_k}{\|\bar{p}_k\|} \right] \qquad (6.11a)$$

and

$$\bar{P}_k = \left[\frac{\bar{p}_1}{\|\bar{p}_1\|}, \frac{\bar{p}_2}{\|\bar{p}_2\|}, \ldots, \frac{\bar{p}_k}{\|\bar{p}_k\|} \right] = AP_k \qquad (6.11b)$$

then by virtue of (5.1d,e) there exists an upper bidiagonal matrix, L_k^T say, such that

$$R_k = P_k L_k^T \qquad (6.12a)$$

and

$$\bar{R}_k = \bar{P}_k L_k^T . \qquad (6.12b)$$

Then it follows using the biorthogonality conditions that

$$\bar{R}_k^T A R_k = L_k \bar{P}_k^T A P_k L_k^T = L_k L_k^T . \qquad (6.13)$$

But from (6.10),

$$\begin{aligned} \bar{R}_k^T A R_k &= \mathcal{D}_k^{\frac{1}{2}} \tilde{\mathcal{L}}_k^T \mathcal{R}_{k+1}^T A \mathcal{R}_k \mathcal{L}_k^{-T} \mathcal{D}_k^{-\frac{1}{2}} \\ &= \mathcal{D}_k^{\frac{1}{2}} \tilde{\mathcal{L}}_k^T \tilde{T}_k \mathcal{L}_k^{-T} \mathcal{D}_k^{-\frac{1}{2}} \\ &= \mathcal{D}_k^{\frac{1}{2}} \tilde{\mathcal{L}}_k^T \tilde{\mathcal{L}}_k \mathcal{D}_k^{\frac{1}{2}} = B_k^T B_k \end{aligned} \qquad (6.14)$$

from (6.7). Since Choleski factors are unique, it follows that

$$L_k^T = B_k . \qquad (6.15)$$

Thus, using (6.12b), (6.10b) and (6.7) in turn,

$$\bar{P}_k = \bar{R}_k B_k^{-1} = \mathcal{R}_{k+1} \tilde{\mathcal{L}}_k \mathcal{D}_k^{\frac{1}{2}} B_k^{-1} = \mathcal{R}_{k+1} Q_{k+1} \binom{I}{0} = W_k$$

showing the equivalence of the directions $\bar{p}_k$ in biconjugate gradients and the directions w_k in the Paige and Saunders method. Since the methods using the sequence $\hat{x}_k$ in section 5 start and terminate at the same points as the Paige and Saunders method, it follows that $\hat{x}_k = x_k^{PS}$ for all k.

Another feature which follows from equation (6.10a) is that by virtue of (2.11)

$$\mathcal{P}_k = R_k \mathcal{D}_k^{\frac{1}{2}} , \qquad (6.16)$$

that is to say, the residuals r_k in the minimum residual algorithm are proportioned to the search directions p_k^{CG} in the conjugate gradient algorithm (2.1).

Finally two collinearity properties between vectors in different methods are described. The first is that both x_k, x_{k+1} and x_{k+1}^{CG} are collinear in the direction p_k of the minimum residual algorithm. This result is given for example by Hestenes and Stiefel [2] but is readily deduced inductively using the fact that both r_{k+1} and r_{k+1}^{CG} are orthogonal to $\bar{p}_{k-1}$. The second property is that $\hat{x}_k$, x_k^{CG} and x_{k+1}^{CG} are collinear in the direction p_{k+1}^{CG} of of the conjugate gradient algorithm. This result is given by Paige and Saunders. These properties are illustrated in the figure.

Figure

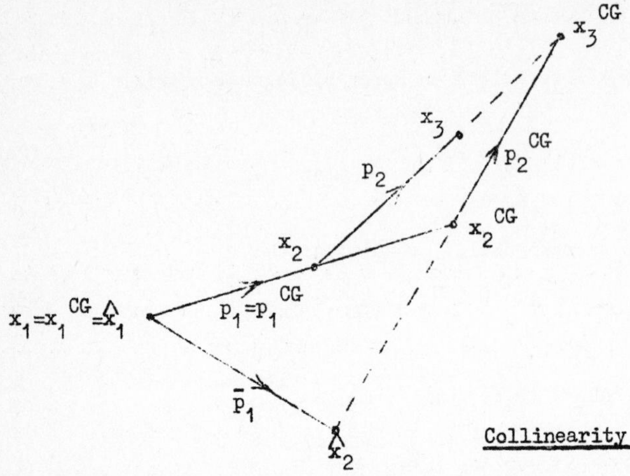

Collinearity properties

Note that $\hat{x}_3$ cannot be drawn on this figure (assuming $m > 2$), but in three dimensions would be at the foot of a perpendicular dropped from $\hat{x}_2$ to the line joining x_3^{CG} and x_4^{CG}.

Collinearity properties can be important in the design of a practical algorithm; for instance it may be possible to recur expressions for $\|r_k^{CG}\|$ or $\|r_k\|$ whilst implementing the orthogonal directions algorithm. Then the algorithm could be terminated if either of $\|r_k^{CG}\|$ and $\|r_k\|$ (in addition to $\|\hat{r}_k\|$) were below a given tolerance. The corresponding value of x_k^{CG} or x_k would then be determined using the appropriate collinearity property.

7. Summary

Various conjugate gradient methods have been examined which can be applied generally to solving symmetric indefinite systems of linear or nonlinear equations. The modification of Luenberger's work has not been followed up because of the potential stability problems. This leaves two methods, the minimum residual algorithm and the orthogonal direction algorithm. Both of these have been shown to be special cases of the biconjugate gradient algorithm. The orthogonal direction algorithm has also been shown to be equivalent to the Paige and Saunders [6] algorithm, although the recurrences given here are quite different, and much more simple. It is thought that in the form given here, the orthogonal direction algorithm has wide potential application, both to linear and nonlinear problems. No substantial numerical experience is yet available, the recurrences merely having been verified on two small problems. However Paige and Saunders

[6] report good although somewhat limited computational experience with their algorithm.

Further work will include trying to apply the orthogonal directions algorithm to large scale partial differential equation problems. However I am conscious that no specific use of the structure in the matrix (1.1) is being used, and it might be profitable to take more account of this. In the context of this paper this might involve other stable reductions of $\tilde{T}_k$ to upper triangular form, or perhaps using a linear combination of p_k and $\bar{p}_k$ as a search direction. There is also a different conjugate gradient algorithm based on the existence of a reduction of A to the form PLQ^T, where P and Q are orthogonal and L is lower bidiagonal. Lawson [4] gives a list of references to this idea. Unfortunately this algorithm requires two multiplications by A per iteration. However it is not simply related to any of the algorithms in his paper, and might therefore be worth considering.

I would finally like to acknowledge the help of Dr T L Freeman and A S M Halliday in carefully reading the manuscript and pointing out a number of mistakes.

References

1. Bunch, J.R., and Parlett, B.N., Direct methods for solving symmetric indefinite systems of linear equations, S.I.A.M. J. Numer. Anal., Vol.8, 1971, pp.639-655.

2. Hestenes, M.R., and Stiefel, E., Methods of Conjugate Gradients for Solving Linear Systems, J. Res. Nat. Bur. Standards, Vol.49, 1952, pp.409-436.

3. Lanczos, C., An Iteration Method for the Solution of the Eigenvalue Problem of Linear Differential and Integral Operators, J. Res. Nat. Bur. Standards, Vol. 45, 1950, pp.255-282.

4. Lawson, C.L., Sparse Matrix Methods Based on Orthogonality and Conjugacy, Jet Propulsion Lab., Tech. Memo. 33-627, 1973.

5. Luenberger, D.G., Hyperbolic pairs in the method of conjugate gradients, S.I.A.M. J. Appl. Math., Vol.17, 1969, pp.1263-1267.

6. Paige, C.C., and Saunders, M.A., Solution of sparse indefinite systems of equations and least squares problems, Stanford University Report, STAN-CS-73-399, 1973.

7. Reid, J.K., On the Method of Conjugate Gradients for the Solution of Large Sparse Systems of Linear Equations, pp.231-254 in Large Sparse Systems of Linear Equations, ed. J.K. Reid, Academic Press, London, 1971.

8. Rutishauser, H., Theory of gradient methods, Chapter 2 of Refined iterative methods for computation of the solution and the eigenvalues of self-adjoint boundary value problems, by M. Engeli, Th. Ginsburg, H. Rutishauser, and E. Stiefel, Birkhaüser, Basel, 1959.

OPTIMAL INTERPOLATION

P.W. Gaffney[+] and M.J.D. Powell

Summary The classical interpolation problem is considered of estimating a function of one variable, $f(.)$, given a number of function values $f(x_i)$, $i=1,2,\ldots,m$. If a bound on $||f^{(k)}(.)||_\infty$ is given also, $k \le m$, then bounds on $f(\xi)$ can be found for any ξ. A method of calculating the closest bounds is described, which is shown to be relevant to the problem of finding the interpolation formula whose error is bounded by the smallest possible multiple of $||f^{(k)}(.)||_\infty$, when $||f^{(k)}(.)||_\infty$ is unknown. This formula is identified and is called the optimal interpolation formula. The corresponding interpolating function is a spline of degree $(k-1)$ with $(m-k)$ knots, so it is very suitable for practical computation.

1. The optimal interpolation problem

To explain the ideas of this paper it is convenient to refer to a sample problem. We use the data given in Table 1, and suppose that we wish to estimate the function value $f(3.5)$ by an interpolation formula whose error is bounded by a multiple of $||f^{(iv)}(.)||_\infty$.

Table 1
Sample data

x	1.0	2.0	3.0	4.0	5.0	6.0
f(x)	-2.0	2.0	1.0	0.5	1.0	-5.0

The most common interpolation method of this type, known as Lagrange interpolation, is obtained by passing a cubic polynomial through four of the data points (see Hildebrand, 1956, for example). Since we are given six data points, there are fifteen different Lagrange interpolation cubic polynomials. It is usually best to make use of the data points that are closest to the point of interpolation. Thus we obtain the formula

$$f(3.5) = -\frac{1}{16}f(2) + \frac{9}{16}f(3) + \frac{9}{16}f(4) - \frac{1}{16}f(5) + \frac{3}{128}f^{(iv)}(\theta)$$

$$= 0.65625 + 0.0234375\, f^{(iv)}(\theta), \qquad (1.1)$$

[+]Present address: Oxford University Computing Laboratory, 19 Parks Road, Oxford.

where θ is in the interval $2 \le \theta \le 5$. We will find that it is also interesting to consider the cubic polynomial through the last four data points, which gives the formula

$$f(3.5) = \frac{5}{16} f(3) + \frac{15}{16} f(4) - \frac{5}{16} f(5) + \frac{1}{16} f(6) - \frac{5}{128} f^{(iv)}(\theta_1)$$

$$= 0.15625 - 0.0390625 \, f^{(iv)}(\theta_1), \tag{1.2}$$

where $3 \le \theta_1 \le 6$.

To show why formula (1.2) may be useful in addition to formula (1.1), we suppose it is known that the inequality

$$||f^{(iv)}(\cdot)||_\infty \le 10 \tag{1.3}$$

is satisfied. In this case expression (1.1) provides the bounds

$$0.421875 \le f(3.5) \le 0.890625 \tag{1.4}$$

and expression (1.2) provides the bounds

$$-0.234375 \le f(3.5) \le 0.546875. \tag{1.5}$$

Thus we find the inequalities

$$0.421875 \le f(3.5) \le 0.546875, \tag{1.6}$$

and the question arises of obtaining the closest bounds on $f(3.5)$, when $f(\cdot)$ satisfies condition (1.3) and has the function values given in Table 1.

This is an example of the following optimal estimation problem. Given the function values $f(x_i)$ $(1 \le i \le m)$ and a bound

$$||f^{(k)}(\cdot)||_\infty \le M \tag{1.7}$$

on the kth derivative of $f(\cdot)$, $k \le m$, what are the best limits on $f(x)$ for any fixed value of x. The solution to this problem is described in Section 2. It is obtained mainly by following the properties of Chebyshev systems given in the excellent book by Karlin and Studden (1966). However their work is not immediately applicable to the optimal estimation problem, because B-splines provide weak Chebyshev systems rather than Chebyshev systems. Because the implications of this last remark have been analysed thoroughly by Gaffney (1975), the material of Section 2 is entirely descriptive.

To introduce the optimal interpolation problem, we note that equations (1.1) and (1.2) provide the bounds

$$|f(3.5) - 0.65625| \le 0.0234375 \, ||f^{(iv)}(\cdot)||_\infty \tag{1.8}$$

and

$$|f(3.5) - 0.15625| \le 0.0390625 \; ||f^{(iv)}(.)||_\infty . \tag{1.9}$$

There are many other interpolation formulae that also give bounds of the form

$$|f(3.5) - s(3.5)| \le c(3.5) \; ||f^{(iv)}(.)||_\infty , \tag{1.10}$$

where s(3.5) and c(3.5) are calculable numbers, depending on the interpolation formula that is used. Among all the interpolation formulae that use the data of Table 1, there is one that gives the smallest possible value of the factor c(3.5). We call it the optimal interpolation formula.

In general terms the optimal interpolation problem is as follows. Given the function values $f(x_i)$ $(1 \le i \le m)$, and given that $||f^{(k)}(.)||_\infty$ is bounded, $k \le m$, but that the actual value of the bound is unknown, to find functions s(x) and c(x) such that the inequality

$$|f(x) - s(x)| \le c(x)||f^{(k)}(.)||_\infty \tag{1.11}$$

must hold, where, for all x, c(x) is as small as possible. The solution to this problem is given in Section 3, and a method of calculating s(x) is indicated. Our method of obtaining the solution is derived from the work of Section 2. However a completely different method was found recently by Miccheli, Rivlin and Winograd (1975), which makes use of perfect splines and Rolle's theorem in an elegant way.

Because the optimal interpolating function s(x) is in fact a spline of degree (k-1) with (m-k) knots, it is very suitable for practical computation. This point and related matters are discussed in Section 4. We note that a less satisfactory optimal interpolating function occurs if $||f^{(k)}(.)||_\infty$ is replaced by $||f^{(k)}(.)||_2$ in inequality (1.11), for in this case c(x) is least when s(x) is a spline of degree (2k-1) with m knots (see Ahlberg, Nilson and Walsh, 1967, for instance).

2. The range of possible values of $f(\xi)$

In order to find the range of possible values of $f(\xi)$ for any fixed ξ when the function values $f(x_i)$ $(i=1,2,...,m)$ and the bound (1.7) are given, $k \le m$, we let Δ be the set of functions $\delta(x)$ such that the equation

$$\delta(x) = f^{(k)}(x) \tag{2.1}$$

is consistent with the data. Therefore each function $\delta(x)$ must satisfy the inequality

$$|\delta(x)| \le M \tag{2.2}$$

and some conditions that are implied by the given function values when $m>k$. These conditions can be expressed in terms of the kth order divided differences

$$f(x_j, x_{j+1}, \ldots, x_{j+k}) = c_j, \quad 1 \le j \le m-k, \tag{2.3}$$

say, whose numerical values are computable from the data. Specifically $\delta(x)$ must satisfy the equations

$$\int \delta(x)\, B_j(x)\, dx = c_j, \quad 1 \le j \le m-k, \tag{2.4}$$

where $B_j(x)$ is a B-spline whose knots are $x_j, x_{j+1}, \ldots, x_{j+k}$ (Schoenberg, 1964), and where the range of integration is the range of x for which $B_j(x)$ is positive. Expressions (2.2) and (2.4) are necessary and sufficient conditions for $\delta(x)$ to belong to Δ.

Sometimes Δ is empty, the simplest example being when M is zero and at least one c_j is non-zero. In this case the data is inconsistent. However, when the data points $x_i (i=1,2,\ldots,m)$ are distinct which we assume is true, then Δ is not empty provided that M is sufficiently large. The description of this section assumes that M is greater than the least value of $||f^{(k)}(\cdot)||_\infty$ that is consistent with the data.

To relate the required range of $f(\xi)$ to the elements of Δ, we suppose that ξ is not a data point, and we define the function $B_\xi(x)$ by the equation

$$\int f^{(k)}(x)\, B_\xi(x)\, dx = f(x_1, x_2, \ldots, x_k, \xi). \tag{2.5}$$

Thus $B_\xi(x)$ is a B-spline whose knots are $x_1, x_2, \ldots, x_k$ and ξ. If the left-hand side of equation (2.5) is known, then the value of $f(\xi)$ can be calculated by using the given function values $f(x_i)$, $i=1,2,\ldots,k$. Therefore finding the range of possible values of $f(\xi)$ is equivalent to obtaining the range of the expression

$$\int \delta(x)\, B_\xi(x)\, dx, \quad \delta(x) \in \Delta . \tag{2.6}$$

We have now expressed the problem in a way that yields to the method used in Section VIII.8 of Karlin and Studden (1966). In order to apply their method we depend on the fact that the functions $B_j(x)$ $(j=1,2,\ldots,m-k)$ of equation (2.4) form a weak Chebyshev set, and that this set augmented by $B_\xi(x)$ is also a weak Chebyshev set. It shows that an extreme value of expression (2.6) is obtained when $\delta(x) \in \Delta$ is a "bang-bang function" having (m-k) sign changes. In other words, if we can find values of $u_1, u_2, \ldots, u_{m-k}$ such that the function

$$\delta_u(x) = \begin{cases} M, & x < u_1, \\ (-1)^t M, & u_t \le x < u_{t+1}, \\ (-1)^{m-k} M, & x \ge u_{m-k}, \end{cases} \tag{2.7}$$

satisfies conditions (2.4), then $\int \delta_u(x)\, B_\xi(x)\, dx$ is an extreme value of expression (2.6). Similarly, if we can also find values of $v_1, v_2, \ldots, v_{m-k}$ such that the function

$$\delta_\ell(x) = \begin{cases} -M, & x < v_1 , \\ (-1)^{t+1}M, & v_t \le x < v_{t+1}, \\ (-1)^{m-k+1}M, & x \ge v_{m-k}, \end{cases} \tag{2.8}$$

satisfies condition (2.4), then $\int \delta_\ell(x)\, B_\xi(x)\, dx$ is the other extreme value of expression (2.6). Therefore our problem reduces to the calculation of the quantities $(u_1,u_2,\ldots,u_{m-k})$ and $(v_1,v_2,\ldots,v_{m-k})$. Because we assume that M is larger than the least value of $||f^{(k)}(\cdot)||_\infty$ that is consistent with the given function values, this calculation has a unique solution (Gaffney, 1975).

One very useful property of $(u_1,u_2,\ldots,u_{m-k})$ and $(v_1,v_2,\ldots,v_{m-k})$ is that they do not depend on ξ. Therefore the functions $\delta_u(x)$ and $\delta_\ell(x)$ provide the range of possible values of $f(\xi)$ for any fixed ξ.

The quantities $u_1,u_2,\ldots,u_{m-k}$ may be calculated by solving a system of nonlinear equations, which are the conditions

$$\int \delta_u(x)\, B_j(x)\, dx = c_j, \quad j=1,2,\ldots,m-k, \tag{2.9}$$

where $\delta_u(x)$ is defined by equation (2.7). Because the bang-bang structure of $\delta_u(x)$ gives the derivative

$$\frac{d}{du_t}\{ \int \delta_u(x)\, B_j(x)\, dx - c_j\} = 2M(-1)^{t+1}\, B_j(u_t), \tag{2.10}$$

the Jacobian matrix of the nonlinear system is easy to evaluate and it has a band structure, which is very helpful in practice. The calculation of $v_1,v_2,\ldots,v_{m-k}$ is similar. Further details are given by Gaffney (1975).

Thus we find the functions $\delta_u(x)$ and $\delta_\ell(x)$ such that the equations $\delta_u(x) = f^{(k)}(x)$ and $\delta_\ell(x) = f^{(k)}(x)$ are each consistent with the data, and which yield the range of $f(\xi)$ for any fixed ξ. Upper and lower limits on $f(x)$ are obtained by integrating $\delta_u(x)$ and $\delta_\ell(x)$ k times, where the constants of integration are chosen so that these integrals agree with $f(x)$ at k of the data points. Agreement is obtained at the other data points because conditions (2.3) and (2.4) are satisfied. We let $u(x,M)$ and $\ell(x,M)$ denote the appropriate k-fold integrals of $\delta_u(x)$ and $\delta_\ell(x)$. The bounds

$$\min[u(\xi,M),\ \ell(\xi,M)] \le f(\xi) \le \max[u(\xi,M),\ \ell(\xi,M)] \tag{2.11}$$

are the closest bounds on $f(\xi)$ for all ξ.

Because $\delta_u(x)$ and $\delta_\ell(x)$ are each bang-bang functions having (m-k) sign changes, the functions $u(x,M)$ and $\ell(x,M)$ are each perfect splines of degree k having (m-k) knots.

For the data of Table 1 when M=10, the method of this section provides the optimal bounds

$$0.4286 \leq f(3.5) \leq 0.5431, \tag{2.12}$$

which are only a little narrower than expression (1.6). For larger values of M the optimal bounds give a greater improvement over those that can be obtained by Lagrange interpolation. For example when M = 500 the optimal bounds are $-6.9765 \leq f(3.5) \leq 8.0019$, while the Lagrangian method provides the bounds $-11.0625 \leq f(3.5) \leq 12.3750$. The differences between the two methods are usually greater when there are more data points.

3. Optimal Interpolation

We now turn to the problem of identifying the interpolating function s(x) through the data $f(x_i)$, i=1,2,...,m, such that the error bound

$$|f(x) - s(x)| \leq c(x) \, ||f^{(k)}(\cdot)||_\infty \tag{3.1}$$

holds and c(x) is as small as possible, where $k \leq m$ and the actual value of $||f^{(k)}(\cdot)||_\infty$ is unknown. The work of Section 2, in particular inequality (2.11), shows that the best bound of this type when the value

$$||f^{(k)}(\cdot)||_\infty = M \tag{3.2}$$

is given is the expression

$$|f(x) - s(x,M)| \leq c(x,M) \, ||f^{(k)}(\cdot)||_\infty , \tag{3.3}$$

where s(x,M) and c(x,M) are the functions

$$s(x,M) = \tfrac{1}{2}[u(x,M) + \ell(x,M)] \tag{3.4}$$

and

$$c(x,M) = \tfrac{1}{2}|u(x,M) - \ell(x,M)|/M. \tag{3.5}$$

Note that the interpolating function s(x,M) is a spline of degree k. Because the bound (3.3) is the best that can be achieved when equation (3.2) holds, and because equation (3.1) has to be satisfied for all values of M, we deduce the inequality

$$c(x) \geq c(x,M). \tag{3.6}$$

The main result of this section is that the required functions s(x) and c(x) are the limits of s(x,M) and c(x,M) as M tends to infinity. It is a consequence of the following theorem.

Theorem The inequality (3.3) is satisfied, not only when equation (3.2) holds, but also when $||f^{(k)}(\cdot)||_\infty$ is less than M.

Proof We let η be the positive number

$$\eta = M - ||f^{(k)}(\cdot)||_\infty , \tag{3.7}$$

and as in Section 2 we let $\delta_u(x)$ and $\delta_\ell(x)$ be the functions whose k-fold integrals are $u(x,M)$ and $\ell(x,M)$. It follows from the properties of $\delta_u(x)$ and $\delta_\ell(x)$ that the function

$$\phi(x) = \delta_u(x) - \delta_\ell(x) \tag{3.8}$$

satisfies the conditions

$$||\phi(\cdot)||_\infty \le 2M \tag{3.9}$$

and

$$\int B_j(x)\, \phi(x)\, dx = 0, \quad j=1,2,\ldots,m-k. \tag{3.10}$$

Therefore the function

$$\psi(x) = f^{(k)}(x) \pm (\eta/2M)\, \phi(x), \tag{3.11}$$

where we leave open the choice of the $\pm$ sign, is in the set Δ defined at the beginning of Section 2. It follows that a k-fold integral of $\psi(x)$, namely the function

$$\sigma(x) = f(x) \pm (\eta/2M)\, \{u(x,M) - \ell(x,M)\} , \tag{3.12}$$

is between the functions $u(x,M)$ and $\ell(x,M)$. Therefore, letting ξ be any fixed value of x, we have the condition

$$|\sigma(\xi) - s(\xi,M)| \le M\, c(\xi,M). \tag{3.13}$$

Substituting the definition of $\sigma(\xi)$ and using equation (3.5) gives the bound

$$|f(\xi) \pm [\pm\, \eta c(\xi,M)] - s(\xi,M)| \le M\, c(\xi,M), \tag{3.14}$$

where the $\pm$ sign inside the square brackets is the sign of $\{u(\xi,M) - \ell(\xi,M)\}$, and where the other $\pm$ sign is obtained from equation (3.11) and is at our disposal. Therefore one choice of this $\pm$ sign, depending on ξ, provides the inequality

$$\begin{aligned} |f(\xi) - s(\xi,M)| &\le (M-\eta)\, c(\xi,M) \\ &= c(\xi,M)\, ||f^{(k)}(\cdot)||_\infty . \end{aligned} \tag{3.15}$$

Since the choice of ξ is arbitrary, inequality (3.3) is satisfied for all values of x, which completes the proof of the theorem.

The theorem shows that inequality (3.3) is valid when $||f^{(k)}(\cdot)||_\infty$ is unknown provided that M is large enough. Therefore, if the limits

$$\left.\begin{aligned} \bar{s}(x) &= \lim_{M\to\infty} s(x,M) \\ \bar{c}(x) &= \lim_{M\to\infty} c(x,M) \end{aligned}\right\} \tag{3.16}$$

exist, the inequality

$$|f(x) - \bar{s}(x)| \le \bar{c}(x)\ ||f^{(k)}(\cdot)||_\infty \tag{3.17}$$

holds whenever $||f^{(k)}(\cdot)||_\infty$ is bounded. Now the function $\bar{s}(x)$ is the optimal interpolating function if and only if $\bar{c}(x)$ is the least value of $c(x)$ that can occur in the bound (3.1). Expression (3.6) shows that the required $c(x)$ is bounded below by $c(x,M)$ for all M. It follows from the definition (3.16) that we have found the optimal interpolating function.

Because the limits of expression (3.16) are considered in detail by Gaffney (1975), we now describe the main properties of $\bar{c}(x)$ and $\bar{s}(x)$ without giving much proof.

The functions $\bar{c}(x)$ and $\bar{s}(x)$ are bounded because inequality (3.17) is at least as good as any bound of the form (3.1) that is obtained by the Lagrange interpolation method.

By dividing the equations (2.9) by the factor M, we find that as $M \to \infty$ the points $(u_1, u_2, \ldots, u_{m-k})$ are defined by the conditions

$$\int h(x)\, B_j(x)\, dx = 0, \quad j=1,2,\ldots,m-k, \tag{3.18}$$

where $h(x)$ is the function

$$h(x) = \begin{cases} 1, & x < u_1, \\ (-1)^t, & u_t \le x \le u_{t+1}, \\ (-1)^{m-k}, & x \ge u_{m-k}. \end{cases} \tag{3.19}$$

Thus the limit of $(u_1, u_2, \ldots, u_{m-k})$ can be calculated, and we call it $(u_1^*, u_2^*, \ldots, u_{m-k}^*)$. Note that it depends on the data points x_i $(i=1,2,\ldots,m)$ but not on the function values $f(x_i)$ $(i=1,2,\ldots,m)$. Because $h(x)$ can be replaced by $-h(x)$ in condition (3.18), it follows that the limits

$$v_t \to u_t^*, \quad t=1,2,\ldots,m-k, \tag{3.20}$$

are also obtained as $M \to \infty$. Thus for each t the difference $(u_t - v_t)$ tends to zero, and $\bar{c}(x)$ is the modulus of the k-fold integral of $h(x)$, where the constants of integration are chosen so that the k-fold indefinite integral is zero at the data points. The fact that $\bar{c}(x)$ is the modulus of a perfect spline was also discovered by Micchelli, Rivlin and Winograd (1975).

Because the definitions of $s(x,M)$, $u(x,M)$ and $\ell(x,M)$ give the derivative

$$\begin{aligned} s^{(k)}(x,M) &= \tfrac{1}{2}[u^{(k)}(x,M) + \ell^{(k)}(x,M)] \\ &= \tfrac{1}{2}[\delta_u(x) + \delta_\ell(x)]\ , \end{aligned} \tag{3.21}$$

it follows from equations (2.7) and (2.8) that $s^{(k)}(x,M)$ is zero if x is in the intersection of (u_t,u_{t+1}) and (v_t,v_{t+1}) for any t. Therefore the convergence of (u_t-v_t) to zero implies that $\overline{s}^{(k)}(x)$ is zero unless x is in the point set $\{u_t^*;\ t=1,2,\dots,m-k\}$. Therefore the optimal interpolating function is a piecewise polynomial, having (m-k) joins, the degree of each polynomial being at most (k-1).

In addition to the limit (3.20), it can be proved that for each t $(1 \le t \le m-k)$ the quantity $M(u_t-v_t)$ tends to a limit as $M \to \infty$ (Gaffney, 1975). It follows that $\overline{s}^{(k)}(x)$ can be regarded as a sum of Dirac delta-functions whose weights are at u_t^* $(1 \le t \le m-k)$. Thus $\overline{s}(x)$ is a spline of degree (k-1).

To calculate the optimal interpolating function, one may first obtain the knots u_t^* $(t=1,2,\dots,m-k)$ from equations (3.18) and (3.19). The remaining parameters of $\overline{s}(x)$ occur linearly, there are m of them, and they are found by solving the equations

$$\overline{s}(x_i) = f(x_i), \quad i=1,2,\dots,m. \tag{3.22}$$

For the sample problem given in Table 1 the knots of $\overline{s}(x)$ are 2.9492 and 4.0508. The optimal interpolation formula gives the bound

$$|f(3.5) - 0.5127| \le 0.014982\ ||f^{(iv)}(\cdot)||_\infty\ , \tag{3.23}$$

which is tighter than the bound (1.8) that comes from the Lagrange interpolation formula (1.1).

4. Discussion

Because the main purpose of the work described is to identify the solution to a basic interpolation problem, it is very fortunate that the resultant interpolation formula is easy to use and that the interpolating function is a spline of degree one less than the derivative of f(x) that occurs in the error term. From a practical point of view we could not wish for a nicer result.

The function $\overline{s}(x)$ is a spline not because the functions $B_j(x)$ in equation (2.4) are splines, but because of the bang-bang properties of $\delta_u(x)$ and $\delta_\ell(x)$ shown in equations (2.7) and (2.8). If the data is changed from function values to some other linear functionals of f(x), then we can still find conditions on $f^{(k)}(x)$ that are analogous to expression (2.4), but in general the kernel functions $B_j(x)$ are not splines. However, provided that the kernel functions satisfy some Chebyshev set conditions, the bang-bang properties of $\delta_u(x)$ and $\delta_\ell(x)$ are retained. Therefore the natural occurrence of splines is not restricted to the interpolation problem.

One curious feature of the optimal interpolation method is that, for a given set of data points x_i $(i=1,2,\dots,m)$, the interpolation operator is linear. This is because the positions of the knots of $\overline{s}(x)$ do not depend on the function values $f(x_i)$ $(i=1,2,\dots,m)$. It is unusual for a linear operator to come from a theory that uses the maximum norm.

We have applied the method to rather few test problems, and the results so far have been entirely satisfactory, in the sense that the calculated functions $\bar{s}(x)$ are like ones that might be estimated by an experienced computer user. Therefore a case can be made for ignoring the origin of the optimal interpolation method, and instead regarding it as no more than an automatic method for passing a function through a sequence of data points. This may prove to be the main application, for the user has to specify only the data points and the degree of the interpolating spline, and then the remainder of the calculation, including the choice of knot positions, is done automatically.

Acknowledgement

Part of this work was supported by a Science Research Council grant, number B/73/794.

References

Ahlberg, J.H., Nilson, E.N. and Walsh, J.L. (1967) "The theory of splines and their applications", Academic Press, New York.

Gaffney, P.W. (1975) "Optimal Interpolation", D.Phil thesis, University of Oxford.

Hildebrand, F.B. (1956) "Introduction to numerical analysis", McGraw-Hill Inc., New York.

Karlin, S. and Studden, W.J. (1966) "Tchebycheff systems: with applications in analysis and statistics", Interscience Publishers, New York.

Micchelli, C.A., Rivlin, T.J. and Winograd, S. (1975) "The optimal recovery of smooth functions", manuscript, IBM Research Laboratories, Yorktown Heights.

Schoenberg, I.J. (1964) "On interpolation by spline functions and its minimal properties", from "On Approximation Theory", eds. P.L. Butzer and J. Korevaar, Birkhaüser Verlag.

Advances in Chebyshev Quadrature*

Walter Gautschi

1. Introduction

Let $d\mu(x)$ be a positive measure on the interval (a,b) admitting finite moments of all orders,

$$(1) \qquad \mu_r = \int_a^b x^r d\mu(x) < \infty, \qquad r = 0,1,2,\dots .$$

We consider quadrature rules of the type

$$(2) \qquad \int_a^b f(x)d\mu(x) = \sum_{k=1}^{n} \gamma_k^{(n)} f(x_k^{(n)}) + R_n(f)$$

having equal weights

$$(3) \qquad \gamma_1^{(n)} = \gamma_2^{(n)} = \dots = \gamma_n^{(n)} .$$

Equally-weighted quadrature sums have the property of minimizing the effect of random errors in the function values $f(x_k^{(n)})$, which may be a useful feature if these errors are considerably larger than the truncation error $|R_n(f)|$. Another, though minor, advantage of equal coefficients results from the fact that only one multiplication is required, as opposed to n, to evaluate the quadrature sum in (2) (not counting the work in evaluating f).

To be widely useful, quadrature rules of the type (2), (3) should have real distinct nodes $x_k^{(n)}$, preferably all located in (a,b). In addition, they should be reasonably accurate. We say that (2), (3) is a Chebyshev quadrature rule, if all nodes are real and if the formula has algebraic degree of exactness n, i.e.,

$$(4) \qquad R_n(f) = 0, \quad \text{all } f \in \mathbb{P}_n .$$

($\mathbb{P}_n$ denotes the class of polynomials of degree $\leq n$.) Letting $f \equiv 1$ in (2) then gives immediately

$$(5) \qquad \gamma_k^{(n)} = \frac{\mu_0}{n}, \qquad k = 1,2,\dots,n .$$

We call (2), (5) a Chebyshev quadrature rule in the strict sense, if (4) holds and the nodes $x_k^{(n)}$ are not only real, but pairwise distinct and all contained in (a,b).

*This work was supported in part by the National Science Foundation under grant GP-36557.

Any quadrature rule (2), (3), on the other hand, with only real nodes, will be referred to as a Chebyshev-type quadrature formula. Such a quadrature rule, therefore, need not have algebraic degree of exactness n, in fact, need not even integrate constants exactly, and is permitted to have repeated nodes, i.e., $x_k^{(n)} = x_l^{(n)}$ for some $k \neq l$.

The monic polynomial of degree n whose zeros are $x_k^{(n)}$, $k = 1,2,\ldots,n$, will be denoted, throughout, by $p_n(x; d\mu)$,

$$(6) \qquad p_n(x; d\mu) = \prod_{k=1}^{n} (x - x_k^{(n)}) = x^n + a_1 x^{n-1} + \cdots + a_n .$$

Requirements (4) and (5) uniquely determine the polynomial $p_n(x; d\mu)$ (cf. §3.2). We may say, therefore, alternatively, that (2), (3) constitutes a Chebyshev quadrature rule if and only if the polynomial $p_n(x; d\mu)$ has only real zeros. We shall have occasion to consider also the polynomials which are orthogonal with respect to the measure $d\mu(x)$; these will be denoted by $\pi_n(x; d\mu)$, $n = 0,1,2,\ldots$.

The study of Chebyshev quadratures began in 1874 with a classical memoir of Chebyshev (Chebyshev [1874]). Important progress has subsequently been made by Bernstein [1937], [1938], Ullman [1966], Geronimus [1946], [1969], and others. Apart from a brief review in Wilf [1967], and traditional treatments in textbooks, no comprehensive account seems to be available. In the following we attempt to review recent advances in this field, covering roughly the period 1945-1975.

2. The classical Chebyshev quadrature formula

2.1 Bernstein's result. The quadrature formula §1(2), (4), (5), in which $d\mu(x) = dx$ on $[-1,1]$, will be referred to as the classical Chebyshev quadrature formula. It was computed by Chebyshev for $n = 2,3,\ldots,7$ and found, in these cases, to have only real nodes, all contained in $[-1,1]$. Radau [1880b] adds to this the case $n=9$, which also yields real nodes, but notes that the formula with $n=8$ involves complex nodes. It took some fifty years after that, until Bernstein, by extremely ingenious arguments, succeeded in proving that the formulas found by Chebyshev and Radau are in fact the only ones which have all nodes real. If $n > 9$, and $n=8$, the polynomial $p_n(x; d\mu)$ necessarily has complex zeros. A simplified version of Bernstein's proof is given by Krylov [1957] (and is also reproduced in Krylov [1962, p.192ff]). Kahaner [1969] interprets the presence of complex nodes as the result of a conflict between the equal coefficients requirement and the polynomial exactness requirement, the former tending to impose a uniform distribution on the real zeros, the latter a non-uniform distribution (as the zeros of orthogonal polynomials), when $n \to \infty$.

There is a fair amount of numerical information available on classical

Chebyshev quadrature. Salzer [1947] exhibits the polynomials $p_n(x; d\mu)$ in exact rational form for $n = 1(1)12$, and also gives the zeros to 10 decimal places for $n = 2(1)7$ and $n = 9$. (The latter are reprinted in Abramowitz and Stegun [1964, p.920].) On the microfiche addendum to Kahaner [1971] the zeros of $p_n(x; d\mu)$, including the complex ones, are tabulated to 14 decimals for $2 \leq n \leq 47$.

2.2 Geometry of the zeros of $p_n(x; d\mu)$. The distribution of the zeros of $p_n(x; d\mu)$ in the complex plane, when $d\mu(x) = dx$ on $[-1,1]$, is studied in detail by Kuzmin [1938]. It turns out that for large n, all zeros (except the zero at the origin, when n is odd) accumulate near the curve

$$(1) \qquad \omega(z) = \omega(1) , \qquad \omega(z) = \int_{-1}^{1} \ln|z-t|dt ,$$

familiar from potential theory. (This is an eye-shaped curve, centred at the origin, which intersects the real axis at ± 1, and the imaginary axis at about $\pm .52$.) More precisely, Kuzmin shows that for n sufficiently large and $h = \sqrt{\ln n}/n$, all zeros of $p_n(z; d\mu)$ (with the exception noted) are either located inside the circles about ± 1, with radii $3h$, or in the narrow band bounded by the curves $\omega(z) = \omega(1)$ and $\omega(z) = \omega(1-12h)$. The zeros thus approach the logarithmic potential curve (1) from the inside. The case $n = 20$ is depicted in Kahaner [1971]. Kuzmin also proves that the number of real zeros of $p_n(z; d\mu)$ is $O(\ln n)$ as $n \to \infty$. Additional properties of the zeros can be found in Mayot [1950] and Kahaner [1971].

3. Mathematical techniques

A number of analytic tools have been developed to deal with the construction of Chebyshev quadratures, or with proofs of nonexistence. We briefly review four of them, and illustrate some by examples.

3.1 Chebyshev's method. This is the method used by Chebyshev in his original memoir (Chebyshev [1874]). The polynomial $p_n(z; d\mu)$ is represented explicitly in the form

$$(1) \qquad \begin{aligned} p_n(z; d\mu) &= E\{\exp(\frac{n}{\mu_0} \int_a^b \ln(z-x)d\mu(x))\} \\ &= E\{z^n \exp(-\frac{n}{\mu_0} \sum_{k=1}^{\infty} \frac{\mu_k}{kz^k})\} , \end{aligned}$$

where $E\{\cdot\}$ denotes the polynomial part of $\{\ \}$. Based on this formula, Chebyshev computes his original quadrature rules (with $d\mu(x) = dx$) for $n = 2,3,\ldots,7$.

In the case of $d\mu(x) = (1-x^2)^{-\frac{1}{2}}$ on $[-1,1]$, formula (1) gives

$$p_n(z;\, d\mu) = E\{(\frac{z+\sqrt{z^2-1}}{2})^n\}$$

$$= E\{(\frac{z+\sqrt{z^2-1}}{2})^n + (\frac{z-\sqrt{z^2-1}}{2})^n\} = \frac{1}{2^{n-1}} T_n(z) ,$$

where $T_n(z)$ is (what is now called) the Chebyshev polynomial of the first kind. In this way, Chebyshev recovers the classical Gauss-type quadrature rule

$$(2) \qquad \int_{-1}^{1} f(x)(1-x^2)^{-\frac{1}{2}}dx = \frac{\pi}{n} \sum_{k=1}^{n} f(x_k^{(n)}) + R_n(f), \qquad x_k^{(n)} = \cos(\frac{2k-1}{2n}\pi) ,$$

which he ascribes to Hermite.

An interesting recent extension of (2) is due to Ullman [1966a,b], who considers

$$(3) \qquad d\mu(x) = (1-x^2)^{-\frac{1}{2}}(1+ax)(1+a^2+2ax)^{-1} \quad \text{on } [-1,1], \qquad -1 < a < 1 ,$$

and finds that

$$z \exp(-\frac{1}{\mu_0} \sum_{k=1}^{\infty} \frac{\mu_k}{kz^k}) = \tfrac{1}{2}(z+\sqrt{z^2-1}+a), \qquad |z| > 1 .$$

Application of (1) thus gives

$$p_n(z;\, d\mu) = E\{(\frac{z+\sqrt{z^2-1}+a}{2})^n\}$$

$$= E\{(\frac{z+\sqrt{z^2-1}+a}{2})^n + (\frac{z-\sqrt{z^2-1}+a}{2})^n - (\frac{a}{2})^n\}$$

$$= \frac{1}{2^{n-1}} T_n^{(a)}(z) ,$$

where $T_n^{(a)}(z)$ is a polynomial of degree n generalizing the Chebyshev polynomial $T_n(z) = T_n^{(o)}(z)$. Ullman shows that $T_n^{(a)}(z)$, for each n, has only real zeros, whenever $-\frac{1}{2} < a < \frac{1}{2}$, thus exhibiting his celebrated example of a weight function, other than the classical one in (2), which admits Chebyshev quadrature for each n. Work along this line is continued by Geronimus [1969] (cf. §5.1).

3.2 <u>Method based on Newton's identity</u>. If the quadrature formula

$$(4) \qquad \int_a^b f(x)d\mu(x) = \frac{\mu_0}{n} \sum_{k=1}^{n} f(x_k^{(n)}) + R_n(f)$$

is to have algebraic degree of exactness n, then the nodes $x_k = x_k^{(n)}$ must satisfy

$$\sum_{k=1}^{n} x_k^r = \frac{n}{\mu_0} \mu_r , \qquad r = 1,2,\ldots,n .$$

Since the moments μ_r are known, these equations determine the first n power sums $s_r = \sum_{k=1}^{n} x_k^r$ of the nodes,

$$s_r = \frac{n}{\mu_0} \mu_r , \qquad r = 1,2,\ldots,n . \tag{5}$$

These in turn determine the coefficients a_k in the polynomial $p_n(x; d\mu)$ by virtue of Newton's identities,

$$\begin{cases} s_1 + a_1 = 0 , \\ s_2 + a_1 s_1 + 2a_2 = 0 , \\ \cdots\cdots\cdots\cdots\cdots \\ s_n + a_1 s_{n-1} + \cdots + a_{n-1} s_1 + n a_n = 0 . \end{cases} \tag{6}$$

The desired nodes x_k in (4) can thus be computed by finding the roots of the algebraic equation $x^n + a_1 x^{n-1} + \cdots + a_n = 0$, where the coefficients a_k are obtained recursively from (6), the s_r being given by (5). This method, first used by Radau [1880a,b], is the one generally applied to compute Chebyshev quadratures. It is extended in Gautschi and Yanagiwara [1974] and Anderson and Gautschi [to appear] to deal with Chebyshev-type quadratures involving repeated nodes.

3.3 Bernstein's method. This is a powerful method for proving nonexistence results. Although Bernstein [1937] deals only with the case of a constant weight function, his method extends immediately to arbitrary measures $d\mu(x)$.

Bernstein's idea is to confront Chebyshev's n-point quadrature formula

$$\int_a^b f(x)d\mu(x) = \frac{\mu_0}{n} \sum_{k=1}^{n} f(x_k^{(n)}) + R_n^{Ch}(f) \tag{7}$$

with the m-point Gauss formula

$$\int_a^b f(x)d\mu(x) = \sum_{r=1}^{m} \lambda_r^{(m)} f(\xi_r^{(m)}) + R_m^G(f) , \tag{8}$$

where $\xi_r^{(m)}$ are the zeros of $\pi_m(x; d\mu)$, assumed in decreasing order,

$$a < \xi_m^{(m)} < \xi_{m-1}^{(m)} < \cdots < \xi_1^{(m)} < b ,$$

and $\lambda_r^{(m)}$ are the corresponding Christoffel numbers. Assuming distinct nodes $x_k^{(n)}$

in (7), all contained in (a,b), and assuming that (7) has polynomial degree of exactness $2m-1$, $m < n$, then Bernstein shows that, necessarily,

$$\frac{\mu_0}{n} \leq \min(\lambda_1^{(m)}, \lambda_m^{(m)}) . \tag{9}$$

This inequality remains valid under the weaker assumption of mere reality of the nodes $x_k^{(n)}$ (Gautschi [1975]).

The road from Bernstein's inequality (9) to nonexistence results is still fraught with considerable technical difficulties, particularly in the case of finite intervals. For measures on infinite intervals, the method appears to apply more easily, as we illustrate with the example of the Laguerre measure, $d\mu(x) = x^{\alpha}e^{-x}dx$. Here, (4) takes the form

$$\int_0^{\infty} f(x)x^{\alpha}e^{-x}dx = \frac{\Gamma(\alpha+1)}{n}\sum_{k=1}^{n} f(x_k^{(n)}) + R_n(f), \qquad \alpha > -1 . \tag{10}$$

The orthonormal polynomials are the normalised Laguerre polynomials,

$$\pi_0(x) = [\Gamma(\alpha+1)]^{-\frac{1}{2}} , \qquad \pi_1(x) = [\Gamma(\alpha+2)]^{-\frac{1}{2}}(\alpha+1-x),\ldots, \tag{11}$$

and

$$\lambda_1^{(m)} = \left(\sum_{k=0}^{m-1} [\pi_k(\xi_1^{(m)})]^2\right)^{-1} \leq \left(\pi_0^2 + [\pi_1(\xi_1^{(m)})]^2\right)^{-1} , \qquad m \geq 2 .$$

Using the known inequality (Krylov [1958])

$$\xi_1^{(m)} > 2m + \alpha - 1 \qquad (m \geq 2,\ \alpha > -1) ,$$

and the explicit expressions in (11), we can further estimate

$$\lambda_1^{(m)} < \frac{\Gamma(\alpha+1)}{1 + 4(m-1)^2/(\alpha+1)} , \qquad m \geq 2 . \tag{12}$$

Now suppose n is even, $n = 2m$ $(m \geq 2)$, and the quadrature rule (10) has degree of exactness n. Then, a fortiori, it has degree of exactness $2m-1$, and hence by Bernstein's inequality,

$$\frac{\Gamma(\alpha+1)}{n} \leq \lambda_1^{(m)} . \tag{13}$$

If (13) is violated, then Chebyshev quadrature in (10) is impossible. By virtue of (12), this will be the case if

$$\frac{1}{n} \geq \frac{1}{1 + 4(m-1)^2/(\alpha+1)} , \qquad m \geq 2 .$$

Since $n = 2m$, the last inequality amounts to $n^2 - (\alpha+5)n + \alpha + 5 \geq 0$, $n \geq 4$, that is, to

(14) $$n \geq \tfrac{1}{2}\{\alpha + 5 + \sqrt{(\alpha+1)(\alpha+5)}\} \quad \text{and} \quad n \geq 4 .$$

For all even values of n satisfying (14), therefore, the Chebyshev formula (10) does not exist. A similar argument applies for n odd, and also for Chebyshev-type quadratures (10) of given degree of exactness $< n$ (Gautschi [1975]).

3.4 <u>Methods based on moment sequences</u>. We already observed in (5) that for (4) to be a Chebyshev quadrature formula, it is necessary and sufficient that the nodes be real and

$$s_r = \frac{n}{\mu_0}\mu_r \quad , \quad r = 1,2,\ldots,n,$$

where $s_r = \sum_{k=0}^{n} x_k^r$ are the power sums in the nodes $x_k = x_k^{(n)}$. Any general property valid for power sums s_r in real variables thus immediately translates into a property for the moments μ_r, $r = 1,2,\ldots,n$, which in turn represents a necessary condition for (4) to be a Chebyshev quadrature rule. Violation of this property implies nonexistence of (4).

One such property, used by Wilf [1961], is Jensen's inequality, which states that for nonnegative numbers, $\xi_k \geq 0$, the quantities $\sigma_r = (\sum_{k=1}^{n} \xi_k^r)^{1/r}$ are nonincreasing in r for $r > 0$, i.e., $\sigma_r \geq \sigma_s$ whenever $0 < r < s$ (Hardy, Littlewood and Pólya [1952, p.28]). Consequently, if all $x_k \geq 0$, then

(15) $$\tau_r = \left(\frac{n}{\mu_0}\mu_r\right)^{\frac{1}{r}} \text{ is nonincreasing for } r = 1,2,\ldots,n,$$

and if all x_k are arbitrary real,

(15*) $$\tau_r^* = \left(\frac{n}{\mu_0}\mu_{2r}\right)^{\frac{1}{r}} \text{ is nonincreasing for } r = 1,2,\ldots,[\tfrac{n}{2}] .$$

Tureckiĭ [1962] and, subsequently, Janovič [1971] and Nutfullin and Janovič [1972] use the more obvious inequalities

$$s_n \leq s_{n-2r}s_{2r} \quad (n \text{ even};\ r = 1,2,\ldots,\tfrac{n}{2}),$$
$$s_{n-1} \leq s_{n-1-2r}s_{2r} \quad (n \text{ odd};\ r = 1,2,\ldots,\tfrac{n-1}{2}),$$

valid for arbitrary real x_k, to obtain the necessary conditions

(16) $$\begin{cases} \dfrac{\mu_n}{n\mu_{n-2r}} \leq \dfrac{\mu_{2r}}{\mu_0} & (n \text{ even};\ r = 1,2,\ldots,\tfrac{n}{2}) , \\[2ex] \dfrac{\mu_{n-1}}{n\mu_{n-1-2r}} \leq \dfrac{\mu_{2r}}{\mu_0} & (n \text{ odd};\ r = 1,2,\ldots,\tfrac{n-1}{2}) . \end{cases}$$

To illustrate (15), consider again the Laguerre weight $x^{\alpha} e^{-x}$, $\alpha > -1$, for which $\mu_r = \Gamma(\alpha+r+1)$, $r = 0,1,2,\ldots$. Requiring nonnegative nodes x_k, we can apply (15), i.e., $\tau_{r-1} \geq \tau_r$ for $2 \leq r \leq n$, which, for $r = n$, gives

$$\left(\frac{n}{\mu_0} \mu_{n-1}\right)^{\frac{1}{n-1}} \geq \left(\frac{n}{\mu_0} \mu_n\right)^{\frac{1}{n}},$$

or, equivalently,

$$\frac{n}{\Gamma(\alpha+1)} \frac{\Gamma(\alpha+n+1)}{(\alpha+n)^n} \geq 1 . \tag{17}$$

Since the left-hand side is asymptotically equal to $\sqrt{2\pi}e^{-\alpha} [\Gamma(\alpha+1)]^{-1} n^{\alpha+3/2} e^{-n}$ as $n \to \infty$, it is clear that (17) will be false for all n sufficiently large, hence Chebyshev quadrature (in the strict sense) not possible.

Similarly, putting $r = 1$ in the first of (16), we find the necessary condition (Tureckiĭ [1962])

$$n^2 - (\alpha^2+\alpha+3)n + \alpha(\alpha-1) \leq 0 , \quad n(\text{even}) \geq 4 ,$$

which leads to a nonexistence result similar to, but not as sharp as, the one obtained in (14) by Bernstein's method.

4. Chebyshev quadrature and Gaussian quadrature

We noted before in §3(2) that the Gauss quadrature formula for $d\mu(x) = (1-x^2)^{-\frac{1}{2}}dx$ on $[-1,1]$ is also a Chebyshev formula. One naturally wonders whether there are other Gauss-type quadrature formulas whose coefficients are all equal. The question was settled negatively at a surprisingly early stage (Posse [1875], Sonin [1887], Krawtchouk [1935], Bailey [1936]). An elegant proof of a rather more far-reaching result is due to Geronimus [1944], [1946] (and also reproduced in Krylov [1962, p.183ff] and Natanson [1965, p.150 ff]).

Let $d\mu(x)$ be a measure which admits a set of orthogonal polynomials, $\{\pi_n(x; d\mu)\}_{n=0}^{\infty}$. Positivity of the measure neednot be assumed. Let $\{\xi_k^{(n)}\}_{k=1}^{n}$ be the zeros of $\pi_n(x; d\mu)$, and consider

$$\int_a^b f(x)d\mu(x) = \frac{\mu_0}{n} \sum_{k=1}^{n} f(\xi_k^{(n)}) + R_n(f) . \tag{1}$$

Then Geronimus proves the following: If for each $n = 1,2,3,\ldots$ we have $R_n(f) = 0$ whenever $f(x) = x$ and $f(x) = x^2$ (if $n > 1$), then $d\mu$ is the Chebyshev measure $d\mu(x) = (1-x^2)^{-\frac{1}{2}}dx$, except for a linear transformation.

The proof can be sketched in a few lines. Introducing the power means in the nodes $\xi_k^{(n)}$,

$$m_r^{(n)} = \left(\frac{1}{n}\sum_{k=1}^{n} [\xi_k^{(n)}]^r\right)^{\frac{1}{r}} ,$$

the hypothesis implies

$$m_1^{(n)} = \frac{\mu_1}{\mu_0} = m_1 , \quad m_2^{(n)} = \left(\frac{\mu_2}{\mu_0}\right)^{\frac{1}{2}} = m_2 ,$$

that is, $m_1^{(n)}$ and $m_2^{(n)}$ are independent of n. Assuming the polynomials $\pi_n(x) = \pi_n(x;\, d\mu)$ normalised to have leading coefficients 1, we have on the one hand, by Newton's identities, that

(2) $$\pi_n(x) = x^n - nm_1x^{n-1} + \frac{n}{2}(nm_1^2 - m_2^2)x^{n-2} - \dots ,$$

and on the other, that

(3) $$\begin{cases} \pi_n(x) = (x-\alpha_n)\pi_{n-1}(x) - \beta_n\pi_{n-2}(x), & n = 1,2,3,\dots, \\ \pi_{-1} = 0 , \quad \pi_0 = 1 \end{cases}$$

for some constants α_n, β_n. Inserting (2) into (3), and comparing coefficients of x^{n-1} and x^{n-2} on either side, gives

$$\alpha_n = \alpha , \qquad n = 1,2,3,\dots,$$

$$\beta_n = \beta , \qquad n = 3,4,5,\dots, \quad \beta_2 = 2\beta ,$$

where

$$\alpha = m_1 , \quad \beta = \tfrac{1}{2}(m_2^2 - m_1^2) .$$

It then follows from (3) that

$$\pi_n(x) = \left(\frac{x-\alpha+\sqrt{(x-\alpha)^2-4\beta}}{2}\right)^n + \left(\frac{x-\alpha-\sqrt{(x-\alpha)^2-4\beta}}{2}\right)^n ,$$

which is essentially the Chebyshev polynomial of the first kind, $T_n(x) = \frac{1}{2}\{(x + \sqrt{x^2-1})^n + (x - \sqrt{x^2-1})^n\}$, except for a linear transformation in the independent variable and a numerical factor (cf. Rivlin [1974, p.5]).

If the measure $d\mu$ is positive, then all $\xi_k^{(n)}$ are real and $|m_1| < m_2$ by the well-known monotonicity of the power mean $m_r^{(n)}$ as a function of r. It then follows that $\beta > 0$.

5. Existence and nonexistence results

Given a positive measure $d\mu(x)$ on (a,b), we say that Chebyshev quadrature is possible for n [in the strict sense] if there exist n real numbers $x_k^{(n)}$ [pairwise distinct in (a,b)] such that

$$(1) \qquad \int_a^b f(x)d\mu(x) = \frac{\mu_0}{n} \sum_{k=1}^{n} f(x_k^{(n)}) + R_n(f)$$

has algebraic degree of exactness n. The finite or infinite sequence $\{n_j\}$ of all those integers $n_j \geq 1$ for which Chebyshev quadrature is possible will be called the T-sequence of $d\mu(x)$. It will be denoted by $T(d\mu)$, or simply by T. We say that the measure $d\mu(x)$ has property T if its T-sequence consists of all natural numbers, property T^{∞}, if its T-sequence is infinite, and property T^{o}, if it is finite. In this terminology, Bernstein's result may be rephrased by saying that the uniform measure $d\mu(x) = dx$ on $[-1,1]$ has the T-sequence $T = \{1,2,3,4,5,6,7,9\}$, hence property T^{o}. The Chebyshev measure $d\mu(x) = (1-x^2)^{-\frac{1}{2}}dx$, on the other hand, has property T.

Bernstein's method, as well as the methods based on moment sequences (cf. §3.3, 3.4) yield necessary conditions for $d\mu(x)$ to have property T^{∞}, hence, by default, also proofs for property T^{o}.

5.1 Measures with property T or T^{∞}. Measures $d\mu(x)$ with property T are rare; in fact, they occur with probability zero, if viewed as moment sequences in appropriate moment spaces (Salkauskas [1975]). Up until Ullman's discovery (cf. §3.1), Chebyshev's measure indeed was the only known measure with property T. Geronimus [1969] continues Ullman's work by first establishing an interesting sufficient condition for Chebyshev quadrature to be possible for n. To describe it, let $d\mu(x) = \omega(x)dx$ on $[-1,1]$, and assume

$$\omega(\cos\theta) = \frac{1}{\pi \sin\theta} \sum_{k=0}^{\infty} a_k \cos k\theta, \quad 0 \leq \theta \leq \pi, \quad a_0 = 1.$$

Define the constants $\{A_m^{(n)}\}_{m=0}^{\infty}$ by

$$\exp\left(-n \sum_{k=1}^{\infty} \frac{a_k}{kz^k}\right) = \sum_{m=0}^{\infty} \frac{A_m^{(n)}}{z^m}, \quad A_0^{(m)} = 1, \quad |z| > 1.$$

Then Chebyshev quadrature is possible for n if the polynomial $\sum_{m=0}^{n-1} A_m^{(n)} z^m + \frac{1}{2}A_n^{(n)} z^n$ has all its zeros in $|z| > 1$. In this case, moreover,

$$2^{n-1} p_n(x;d\mu) = \sum_{m=0}^{n-1} A_m^{(n)} \cos(n-m)\theta + \tfrac{1}{2}A_n^{(n)}, \quad x = \cos\theta.$$

Ullman's measure with property T falls out as a simple example, by taking

$a_k = (-a)^k$. Geronimus also gives several examples of even weight functions $\omega(x)$ admitting Chebyshev quadratures for all even integers $n = 2\nu$. (These automatically have degree of exactness $2\nu + 1$.)

A measure $d\mu(x)$ on $(-\infty,\infty)$ with infinite support (i.e., with positive mass outside of every finite interval) cannot have property T^∞ unless its T-sequence contains very large gaps. For example, if $\{2\nu_j\}$ is the even subsequence of $T(d\mu)$, and m any fixed integer, then one has $\nu_j > \nu_{j-1}^m$ for infinitely many j (Wilf [1961]). Similarly for the odd subsequence. It follows, in particular, that a measure with property T necessarily has finite support. Wilf in fact conjectures that property T^∞ already implies finite support. This, however, is disproved by Ullman [1962], [1963], who in turn poses the question (still open) of formulating criteria in terms of the gaps of an infinite T-sequence, which would allow to discriminate between measures with infinite, and measures with finite, support.

Kahaner and Ullman [1971] establish conditions on the measure $d\mu(x)$ on $(-\infty,\infty)$ which either imply the absence of property T , or property T^o. The conditions involve the limit behaviour (as $n \to \infty$) of certain discrete measures concentrated at the zeros of the orthogonal polynomials $\pi_n(x;d\mu)$.

5.2 Chebyshev quadrature on finite intervals. Soon after Bernstein obtained his classical result, Akhiezer [1937], in a little-known paper, proved that the Jacobi measure $d\mu(x) = (1-x)^\alpha(1+x)^\beta dx$ on $[-1,1]$ has property T^o whenever $-\frac{1}{2} \leq \alpha \leq \frac{1}{2}$, $-\frac{1}{2} \leq \beta \leq \frac{1}{2}$ (excepting $\alpha = \beta = -\frac{1}{2}$). More recently, using Bernstein's method, Gatteschi [1963/64] proves property T^o for all $\alpha = \beta > -\frac{1}{2}$, while Ossicini [1966] extends it to $\alpha > -\frac{1}{2}$, $\beta > -1$, hence, by symmetry, also to $\alpha > -1$, $\beta > -\frac{1}{2}$. In the remaining square $-1 < \alpha \leq -\frac{1}{2}$, $-1 < \beta \leq -\frac{1}{2}$ (with $\alpha = \beta = -\frac{1}{2}$ deleted), the matter appears to be still unsettled.

Greenwood and Danford [1949] consider the integral $\int_0^1 xf(x)dx$ (which amounts to Jacobi's case $\alpha = 0$, $\beta = 1$) and find by computation that Chebyshev quadrature is possible (in the strict sense) if $n = 1,2,3$, but not if $4 \leq n \leq 10$. A similar result is stated in Greenwood, Carnahan and Nolly [1959] for the integral $\int_{-1}^1 x^2f(x)dx$ (which can be reduced to the case $\alpha = 0$, $\beta = \frac{1}{2}$). The exact T-sequence has not been established in either case.

5.3 Chebyshev quadrature on infinite intervals. Computational results of Salzer [1955] suggested that the T-sequence for the Laguerre measure $d\mu(x) = e^{-x}dx$ on $(0,\infty)$, as well as the one for the Hermite measure $d\mu(x) = e^{-x^2}$ on $(-\infty,\infty)$, must be rather short, in fact $T = \{1,2\}$ in the former, and $T = \{1,2,3\}$ in the latter case. This was first proved by Krylov [1958], by an application of Bernstein's method, and again later, independently, by Gatteschi [1964/65]. Burgoyne [1963], unaware of Krylov's result, confirms it up to $n = 50$ by computing the maximum number of nonnegative, resp. real, nodes. For more general Laguerre measures $d\mu(x) = x^{\alpha}e^{-x}$, $\alpha > -1$, property T^{o} is proved by Wilf [1961], Tureckiĭ [1962] and Gautschi [1975], using methods already illustrated in §3.3, 3.4.

Nutfullin and Janovič [1972], using the method of Tureckiĭ, prove property T^{o} for the measures

$$d\mu(x) = (x^{2p+1}/\sinh \pi x)dx , \quad p = 0,1,2,\ldots,$$

$$d\mu(x) = (x^{2p}/\cosh \pi x)dx , \quad p = 0,1,2,\ldots,$$

and

$$d\mu(x) = |x|^{\alpha}e^{-x^2}dx , \quad \alpha > -1 ,$$

all on $(-\infty,\infty)$, and for each give an upper bound for $\max_{n_j\in T(d\mu)} n_j$. They also determine the T-sequence for some of these measures. For example, the first, when $p = 0$, has $T = \{1,2,3\}$, while the last has $T = \{1,2,3\}$ for $-1 < \alpha < 1/3$, $T = \{1,2,3,5\}$ for $1/3 \leq \alpha < 1$, $T = \{1,2,3,4,5\}$ for $1 \leq \alpha \leq 7$, $T = \{1,2,3,4\}$ for $7 < \alpha < 15$. For $\alpha \geq 15$, Chebyshev quadrature is possible when $n = 1,2,3,4,6$ but the exact T-sequence is not known.

Janovič [1971] previously used Tureckiĭ's method to show that a certain measure $d\mu(x)$ on $(0,\infty)$, of interest in the theory of Wiener integrals, has $T = \{1,2\}$.

5.4 Chebyshev-type quadrature. If a measure $d\mu(x)$ has property T^{o}, and $p_n = p_n(d\mu)$ is the maximum degree of exactness of (1), subject to the reality of all nodes, it becomes of interest to determine upper bounds for p_n as $n \to \infty$. In the classical case $d\mu(x) = dx$, Bernstein [1937] already showed that $p_n < 4\sqrt{n}$. For Jacobi measures $d\mu(x) = (1-x)^{\alpha}(1+x)^{\beta}dx$, Costabile [1974] establishes $p_n < c(\alpha,\beta)n^{1/(2\alpha+2)}$, as has previously been found by Meir and Sharma [1967] in the ultraspherical case $\alpha = \beta > -\frac{1}{2}$. In this latter case, Costabile further expresses the constant c explicitly in terms of gamma and Bessel functions. For more general weight functions on $[-1,1]$, having branch point and other singularities at the endpoints, the problem is studied extensively by Geronimus [1969], [1970]. For the Laguerre measure $d\mu(x) = x^{\alpha}e^{-x}dx$, $\alpha > -1$, one finds by Bernstein's method that $p_n < 2 + \sqrt{(\alpha+1)(n-1)}$ if $p_n \geq 3$. Similar bounds hold for symmetric Hermite quadrature rules (Gautschi [1975]; see also Tureckiĭ [1962]).

Chebyshev-type quadratures having degree of exactness 1 always exist. The most familiar example is the composite midpoint rule on $[-1,1]$, with $d\mu(x) = dx$. Another example is the nontrivial extension of the midpoint rule to integrals with arbitrary positive measure, due to Stetter [1968b], which improves upon an earlier extension of Jagerman [1966].

6. Optimal Chebyshev-type quadrature formulas

Only relatively recently have attempts been made to develop Chebyshev-type quadrature formulas in cases where true Chebyshev formulas do not exist. The approach generally consists in replacing the algebraic exactness condition by some optimality condition, unconstrained or constrained. This yields new formulas even in cases where ordinary ones exist.

6.1 Optimal formulas in the sense of Sard. For the classical weight $d\mu(x) = dx$ on $[-1,1]$, consider a Chebyshev-type quadrature formula

$$\int_{-1}^{1} f(x)dx = \frac{2}{n}\sum_{k=1}^{n} f(x_k) + R_n(f) . \tag{1}$$

We require that (1) has polynomial degree of exactness $p < n$,

$$R_n(f) = 0 , \quad \text{all } f \in \mathbb{P}_p , \tag{2}$$

and assume $f \in AC^p[-1,1]$. The remainder $R_n(f)$, as is well known (see, e.g., Sard [1963, p.25]), can then be represented in the form

$$R_n(f) = \int_{-1}^{1} K_p(t)f^{(p+1)}(t)dt ,$$

where $K_p(t) = K_p(t;x_1,x_2,\dots,x_n)$ is the Peano kernel of R_n [cf. §7(3)]. By the Schwarz inequality, therefore,

$$|R_n(f)| \le \gamma_p \|f^{(p+1)}\|_{L_2} , \qquad \gamma_p = \|K_p\|_{L_2} , \tag{3}$$

where $\|u\|_{L_2} = (\int_{-1}^{1}[u(t)]^2dt)^{\frac{1}{2}}$. An optimal Chebyshev-type formula in the sense of Sard is a formula (1), satisfying (2), which minimizes γ_p as a function of $x_1,x_2,\dots,x_n$. Franke [1971] studies such formulas in the cases $p = 0$ and $p = 1$, under the additional assumption of symmetry,

$$x_{n+1-k} + x_k = 0 , \quad k = 1,2,\dots,n. \tag{4}$$

The condition (2) is then automatically satisfied, so that the problem reduces to an unconstrained optimization problem. The solution for $p = 0$, as has been noted previously (Krylov [1962, pp.138-140]), is the composite midpoint rule, for which $\gamma_0 = 2/3n^2$. In the case $p = 1$, numerical answers are given for $2 \le n \le 11$. A similar problem, without the symmetry assumption (4), is considered in Coman [1970].

6.2 Least squares criteria. Instead of minimizing γ_p in (3), we may wish to minimize the errors of (1) which result if the formula is applied to successive monomials. More precisely, given an integer p, with $0 \leq p < n$, and an integer q, with $q \geq n$, or $q = \infty$, we determine the nodes x_k in (1) such that

$$\sum_{j=p+1}^{q} [R_n(x^j)]^2 = \min, \tag{5}$$

subject to

$$R_n(x^j) = 0, \quad j = 1,2,\ldots,p. \tag{6}$$

Symmetry, as in (4), may or may not be imposed.

If $n \leq 7$, or $n = 9$, and $q = n$, Problem (5), (6) is trivially solved by the classical Chebyshev formulas, which drive the objective function in (5) to zero. In the case $p = 0$, and for various choices of q, including $q = \infty$, numerical answers are given by Barnhill, Dennis and Nielson [1969] for $n = 8,10,11$. Kahaner [1970] has analogous results for $q = n$ and $p = n - 1$ or $n - 2$. An interesting (although somewhat counterproductive) feature of this work is the apparent necessity of assuming repeated nodes for the minimization procedures to converge. It is shown in Gautschi and Yanagiwara [1974] that repeated nodes are indeed unavoidable, if $q = n$, whenever the constraints in (6) admit real solutions. The same is proved in Salkauskas [1973] for the case $p = 0$, all nodes being constrained to the interval $[-1,1]$. We conjecture that the same situation prevails for arbitrary $q > n$.

There is computational evidence that the optimal formulas are indeed symmetric, but the question remains open. If we knew that Problem (5), (6) had a unique solution, modulo permutations, symmetry would follow (Gautschi and Yanagiwara [1974]).

6.3 <u>Minimum norm quadratures</u>. A quadrature rule, such as (1), which minimizes the norm of the error functional $R_n(f)$ in some appropriate function space is called a minimum norm quadrature formula. For Chebyshev quadratures, such formulas are studied by Rabinowitz and Richter [1970]. They consider two families of Hilbert spaces. Each space consists of functions which are analytic in an ellipse $\mathcal{E}_\rho$, $\rho > 1$, having foci at ± 1 and semiaxes summing up to ρ. ($\{\mathcal{E}_\rho\}$ is a family of confocal ellipses, which as $\rho \to 1$ shrink to the interval $[-1,1]$, and as $\rho \to \infty$ inflate into progressively more circle-like regions invading the whole complex plane.) The first space, $L^2[\mathcal{E}_\rho]$, contains functions f for which $\iint_{\mathcal{E}_\rho} |f(z)|^2 dxdy < \infty$, and is equipped with the inner product $(f,g) = \iint_{\mathcal{E}_\rho} f(z)\overline{g(z)}dxdy$. The second, $H^2[\mathcal{E}_\rho]$, consists of functions f with $\int_{\partial\mathcal{E}_\rho} |f(z)|^2 |1-z^2|^{-\frac{1}{2}} |dz| < \infty$ and carries the inner product $\int_{\partial\mathcal{E}_\rho} f(z)\overline{g(z)}|1-z^2|^{-\frac{1}{2}}|dz|$.

The norm of $R_n(f)$, in each of these spaces, can be expressed explicitly in terms of the respective orthonormal bases. Thus, in $L^2[\mathcal{E}_\rho]$,

$$\|R_n\| = \frac{4}{\pi} \sum_{j=0}^{\infty} \left[\frac{j+1}{\rho^{2j+2} - \rho^{-2j-2}} R_n(U_j)\right]^2 , \tag{7}$$

where U_j are the Chebyshev polynomials of the second kind, and in $H^2[\mathcal{E}_\rho]$,

$$\|R_n\| = \frac{2}{\pi} {\sum_{j=0}^{\infty}}' \left[\frac{1}{\rho^{2j} + \rho^{-2j}} R_n(T_j)\right]^2 , \tag{8}$$

where T_j are the Chebyshev polynomials of the first kind. (The prime indicates that the term with $j = 0$ is to be halved.) It is shown by Rabinowitz and Richter that there exists a set of nodes x_k in $[-1,1]$ for which (7), and one for which (8), is a minimum, regardless of whether the weight in the quadrature rule is fixed to be $2/n$, as in (1), or whether it is treated as a free parameter. Numerical results given by Rabinowitz and Richter suggest that the optimal nodes are mutually distinct for each $\rho > 1$, but this remains a conjecture.

Rabinowitz and Richter also investigate the behaviour of the optimal Chebyshev-type rules in the limit cases $\rho \to 1$ and $\rho \to \infty$. In the former case, the limit behaviour is somewhat bizarre, and we shall not attempt to describe it here. In the latter case, it follows from (7), (8) that, both in $L^2[\mathcal{E}_\rho]$ and $H^2[\mathcal{E}_\rho]$, the optimal rule must be such that it integrates exactly as many monomials as possible, and gives minimum error for the first monomial which cannot be integrated exactly. Thus,

$$\begin{cases} R_n(x^j) = 0 , \quad j = 0,1,2,\ldots,p, \quad p = \max(= p_n) , \\ |R_n(x^{p+1})| = \min . \end{cases} \tag{9}$$

We call the corresponding quadrature rules briefly <u>E-optimal</u>. Numerical results given by Rabinowitz and Richter for $n = 8,10,11,12,13$ show again the presence of repeated nodes.

6.4 <u>E-optimal quadratures</u>. An algebraic study of E-optimal Chebyshev-type quadrature rules is made in Gautschi and Yanagiwara [1974] for $n = 8,10,11,13$, and in Anderson and Gautschi [to appear] for general n. One of the key results of this work reveals that an E-optimal n-point Chebyshev-type formula can have at most p_n distinct nodes, whenever $p_n < n$. It follows from this immediately that some of the nodes must be repeated. The (generally distinct) p_n optimal nodes are found among the real solutions of systems of algebraic equations of the type

$$\sum_{r=1}^{p} \nu_r x_r^{\,j} = s_j , \qquad j = 1,2,\ldots,p, \tag{10}$$

where ν_r are integers with $\nu_1 + \nu_2 + \dots + \nu_p = n$ and $p = p_n$ an integer generally not known a priori (cf. Eq.(9)). Finding all real solutions of such systems is a challenging computational problem. It is solved in the cited references for $n \leq 17$ by a reduction to single algebraic equations. For other techniques, see also Yanagiwara and Shibata [1974] and Yanagiwara, Fukutake and Shibata [1975]. A summary of results is given below in Table 1, where crosses indicate the availability of E-optimal Chebyshev formulas, zeros the nonexistence of Chebyshev-type quadrature formulas, and question marks unsettled cases.

p \ n	1	2	3	4	5	6	7	8	9	10	11	12	13	14	15	16	17	18	19	20	21	22	23...
2[n/2]+1	X	X	X	X	X	X	X	0	X	0	0	0	0	0	0	0	0	0	0	0	0	0	0...
2[n/2]-1								X		X	X	0	X	0	0	0	0	0	0	0	0	0	0...
2[n/2]-3												X		X	X	0	X	0	0	0	0	0	0...
2[n/2]-5																X		?	?	?	?	0	0...

Table 1. Existence and nonexistence of n-point Chebyshev-type quadrature formulas of degree of exactness p

E-optimal formulas have been obtained also for infinite and semi-infinite intervals involving weight functions of the Hermite and Laguerre type (Anderson and Gautschi [to appear]). The confluence of nodes is rather more severe in these cases. For example, in the Laguerre case, when $3 \leq n \leq 6$, there are only two distinct nodes, one being simple, the other having multiplicity $n - 1$.

7. Error and convergence

7.1 The remainder term. Remainder terms in Chebyshev-type quadratures are generally ignored, except for the classical formulas

$$(1) \qquad \int_{-1}^{1} f(x)dx = \frac{2}{n} \sum_{k=1}^{n} f(x_k^{(n)}) + R_n(f), \qquad n = 1,2,\dots,7,9,$$

and for the Gauss-Chebyshev formula (with $d\mu(x) = (1-x^2)^{-\frac{1}{2}}dx$).

Each of the formulas (1) has polynomial degree of exactness $p = 2[n/2]+1$, that is, $p = n$ if n is odd, and $p = n+1$ if n is even. Assuming $f \in C^{p+1}[-1,1]$, we obtain from Peano's theorem (see, e.g., Davis [1963, p.70])

$$(2) \qquad R_n(f) = \int_{-1}^{1} K_p(t) f^{(p+1)}(t)\,dt ,$$

where $K_p(t)$ is the Peano kernel of the functional $R_n(f)$,

$$(3) \qquad K_p(t) = \frac{1}{p!}\left\{\frac{(1-t)^{p+1}}{p+1} - \frac{2}{n}\sum_{k=1}^{n}(x_k^{(n)} - t)_+^p\right\},$$

with

$$u_+^p = \begin{cases} u^p & \text{if } u \geq 0, \\ 0 & \text{if } u < 0, \end{cases} \qquad p \geq 0 .$$

Ghizzetti and Ossicini [1970], and Kozlovskiĭ [1971], give different proofs of the fact that the Peano kernel is positive,

$$(4) \qquad K_p(t) \geq 0 \quad \text{on} \quad [-1,1] .$$

From (2), it then follows that

$$(5) \qquad R_n(f) = \kappa_n f^{(p+1)}(\tau), \qquad -1 \leq \tau \leq 1 ,$$

where

$$(6) \qquad \kappa_n = \int_{-1}^{1} K_p(t)dt = R_n\left[\frac{x^{p+1}}{(p+1)!}\right], \quad p = 2\left[\frac{n}{2}\right]+1 .$$

Numerical values of the constants κ_n for $n = 1(1)7$ and $n = 9$ can be found in Ghizzetti and Ossicini [1970, pp.129-130]. (They have previously been tabulated by Berezin and Židkov [1965, p.262], but with an incorrect value of κ_9.)

The remainder in the Gauss-Chebyshev quadrature formula has been estimated by a number of writers; see, e.g., Stetter [1968a], Chawla and Jain [1968], Chawla [1969], Riess and Johnson [1969], Chui [1972], Jayarajan [1974].

For E-optimal quadrature rules of the type (1), the remainder $R_n(f)$ is analysed by Anderson [1974].

7.2 Convergence of Chebyshev quadrature formulas. In order to study convergence of the classical Chebyshev quadrature formulas, one must, of course, allow for complex nodes. From the known distribution of the nodes in the complex plane (cf. §2.2) it follows easily from Runge's theorem that convergence is assured for functions which are analytic in a closed domain $\mathcal{D}$ containing the curve of logarithmic potential, §2.2(1), in its interior (Kahaner [1971]). Convergence, in fact, is geometric for $\mathcal{D}$ sufficiently large.

8. Miscellaneous extensions and generalizations of Chebyshev quadrature

There are many variations on the theme of Chebyshev quadrature. A natural thing to try, e.g., is to relax the rigid requirement of equal coefficients and merely seek to minimize some measure of the variance in the coefficients. The problem, first suggested by Ostrowski [1959], is discussed by Kahaner [1969] and

Salkauskas [1971].

A more substantial modification is made by Erdős and Sharma [1965], and Meir and Sharma [1967], who associate equal coefficients only with part of the nodes and leave the coefficients for the remaining nodes, as well as the nodes themselves, variable. Even with this modification, provided the number of variable coefficients is kept fixed, and the polynomial degree of exactness maximized, some of the nodes again turn complex as n, the total number of nodes, becomes large. Erdős and Sharma show this for the measure $d\mu(x) = dx$ on $[-1,1]$, and Meir and Sharma for the ultraspherical measure $d\mu(x) = (1-x^2)^{\alpha}dx$, $\alpha > -\frac{1}{2}$. The maximum polynomial degree of exactness, p_n, subject to the reality of all nodes, when $d\mu(x) = dx$, in fact obeys the law $p_n = O(\sqrt{n})$ familiar from Bernstein's theory of the classical Chebyshev quadratures. For Jacobi measures $d\mu(x) = (1-x)^{\alpha}(1+x)^{\beta}$, $\alpha > -\frac{1}{2}$, $\beta > -1$, Gatteschi, Monegato and Vinardi [to appear] associate variable coefficients with fixed nodes at ± 1, and equal coefficients with the remaining nodes, and for this case, too, establish the impossibility of n-point Chebyshev quadrature for n sufficiently large.

For quadrature sums involving derivative values as well as function values, the natural extension of Chebyshev's problem would be to require equal coefficients for all derivative terms involving the same order derivative. The problem, as far as we know, has not been treated in any great detail, although it is briefly mentioned by Ghizzetti [1954/55] (see also Ghizzetti and Ossicini [1970, p.43ff]).

Chebyshev quadrature rules integrating exactly trigonometric, rather than algebraic, polynomials are considered by Keda [1962] and Rosati [1968]. Rosati includes derivative terms in his quadrature sums.

Equally-weighted quadrature rules for integration in the complex plane are developed by Salzer [1947] in connection with the inversion of Laplace transforms.

An extension of Chebyshev quadrature to double and triple integrals is discussed by Georgiev [1953]. Coman [1970] derives optimal Chebyshev-type formulas in two dimensions.

References

Abramowitz, M., and Stegun, I.A., eds. (1964): Handbook of mathematical functions, Nat. Bur. Standards Appl. Math. Ser. 55 {MR29 #4914}.

Akhiezer, N.I. (1937): On the theorem of S.N.Bernstein concerning the Chebyshev quadrature formula (Ukrainian), Ž. Inst. Mat. Akad. Nauk USSR 3, 75-82 {Zbl. 18, 208}.

Anderson, L.A. (1974): Optimal Chebyshev-type quadrature formulas for various weight functions, Ph.D. Thesis, Purdue University, August 1974.

Anderson, L.A. and Gautschi, W. (to appear): Optimal Chebyshev-type formulas ,Calcolo.

Bailey, R.P. (1936): Convergence of sequences of positive linear functional operations, Duke Math. J. 2, 287-303 {Zbl. 14, 312}.

Barnhill, R.E., Dennis,J.E.,Jr.,and Nielson, G.M. (1969): A new type of Chebyshev quadrature, Math. Comp. 23, 437-441. {MR39 # 3698}.

Berezin, I.S., and Židkov, N.P. (1965): Computing Methods, Vol. I, Pergamon Press, Oxford {MR30 #4372}.

Bernstein, S.N. (1937): Sur les formules de quadrature de Cotes et Tchebycheff, C.R. Acad. Sci. URSS 14, 323-326. [Reprinted in: "Collected Works", Vol.II, Izdat. Akad. Nauk SSSR, Moscow, 1954, pp.200-204 (Russian)] {MR16, 433}.

Bernstein, S.N. (1938): Sur un système d'équations indéterminées, J. Math. Pures Appl. (9) 17, 179-186. [Reprinted in: "Collected Works", Vol.II, Izdat. Akad. Nauk SSSR, Moscow, 1954, pp.236-242 (Russian)] {MR16, 433}

Burgoyne, F.D. (1963): The non-existence of certain Laguerre-Chebyshev quadrature formulas, Math. Comp. 17, 196-197. {MR28, #2634}.

Chawla, M.M. (1969): On Davis' method for the estimation of errors of Gauss-Chebyshev quadratures, SIAM J. Numer. Anal. 6, 108-117. {MR39 #7812}.

Chawla, M.M., and Jain, M.K. (1968): Error estimates for Gauss quadrature formulas for analytic functions, Math. Comp. 22, 82-90. {MR36 #6142}.

Chebyshev, P.L. (1874): Sur les quadratures, J. Math. Pures Appl. (2) 19, 19-34. [Reprinted in: "Oevres", Vol. II, Chelsea, New York, 1962, 165-180] {MR26 #4870}.

Chui, C.K. (1972): Concerning Gaussian-Chebyshev quadrature errors, SIAM J. Numer. Anal. 9, 237-240. {MR46#10177}.

Coman, Gh. (1970): Nouvelles formules de quadrature à coefficients égaux, Mathematica (Cluj) 12 (35), 253-264. {MR48 #12781}.

Costabile, F. (1974): Sulle formule di quadratura di Tschebyscheff, Calcolo 11, 191-200.

Davis, P.J. (1963): Interpolation and Approximation, Blaisdell Publ. Co., New York-Toronto-London. {MR28 #393}.

Erdős, P., and Sharma, A. (1965): On Tchebycheff quadrature, Canad. J. Math. 17, 652-658. {MR31 #3774}

Franke, R. (1971): Best Chebyshev quadratures, Rocky Mountain J. Math. 1, 499-508. {MR43 #6641}

Gatteschi, L. (1963/64): Su di un problema connesso alle formule di quadratura di Tschebyscheff, Univ.e Politec. Torino Rend. Sem. Mat. 23, 75-87. {MR30 #4386}

Gatteschi, L. (1964/65): Sulla non esistenza di certe formule di quadratura, Univ. e Politec. Torino Rend. Sem. Mat. 24, 157-172. {MR32 #4846}

Gatteschi, L., Monegato, G., and Vinardi, G. (to appear): Formule di quadratura quasi gaussiane ed un problema analogo a quello di Tchebycheff, Calcolo.

Gautschi, W. (1975): Nonexistence of Chebyshev-type quadratures on infinite intervals, Math. Comp. 29, 93-99.

Gautschi, W., and Yanagiwara, H. (1974): On Chebyshev-type quadratures, Math. Comp. 28, 125-134. {MR48 #10063}

Georgiev, G. (1953): Formulas of mechanical quadratures with equal coefficients for multiple integrals (Russian), Dokl. Akad. Nauk SSSR 89, 389-392. {MR14, 852}

Geronimus, Ja. L. (1944): On Gauss' and Tchebycheff's quadrature formulas, Bull. Amer. Math. Soc. 50, 217-221. {MR6, 63}

Geronimus, Ja. L. (1946): On Gauss' and Tchebycheff's quadrature formulae, C.R. (Doklady) Acad. Sci. URSS (N.S.) 51, 655-658. {MR10, 37}

Geronimus, Ja. L. (1969): On the Chebyshev quadrature formula (Russian), Izv. Akad. Nauk SSSR Ser. Mat. 33, 1182-1207. [English translation in: Math. USSR-Izv. 3, 1115-1138] {MR41 #4092}

Geronimus, Ja. L. (1970): The order of the degree of precision of Chebyshev's quadrature formula (Russian), Dokl. Akad. Nauk SSSR 190, 263-265. [English translation in: Soviet Math. Dokl. 11, 70-72] {MR41 #7843}

Ghizzetti, A. (1954/55): Sulle formule di quadratura, Rend. Sem. Mat. Fis. Milano 26, 1-16. {MR18, 391}

Ghizzetti, A., and Ossicini, A. (1970): Quadrature Formulae, Academic Press, New York. {MR42 #4012}

Greenwood, R.E., and Danford, M.B. (1949): Numerical integration with a weight function x , J. Math. and Phys. 28, 99-106. {MR11, 57}

Greenwood, R.E., Carnahan, P.D.M., and Nolley, J.W. (1959): Numerical integration formulas for use with weight functions x^2 and $x/\sqrt{1-x^2}$, Math. Tables Aids Comput. 13, 37-40. {MR21 #968}

Hardy, G.H., Littlewood, J.E., and Pólya, G. (1952): Inequalities, 2d ed., Cambridge University Press. {MR 13, 727}

Jagerman, D. (1966): Investigation of a modified mid-point quadrature formula, Math. Comp. 20, 79-89. {MR32 #8499}

Janovič, L.A. (1971): A quadrature formula with equal coefficients for a certain form of the integral (Russian), Dokl. Akad. Nauk BSSR 15, 873-876. {MR44#6150}

Jayarajan, N. (1974): Error estimates for Gauss-Chebyshev and Clenshaw-Curtis quadrature formulas, Calcolo 11, 289-296.

Kahaner, D.K. (1969): On equal and almost equal weight quadrature formulas, SIAM J. Numer. Anal. 6, 551-556. {MR44 #3492}

Kahaner, D.K. (1970): Chebyshev type quadrature formulas, Math. Comp. 24, 571-574. {MR42 #8694}

Kahaner, D.K. (1971): Some polynomials for complex quadrature, Math. Comp. 25, 827-830. {MR45 #7990}

Kahaner, D.K., and Ullman, J.L. (1971): Equal weight quadrature on infinite intervals, SIAM J. Numer. Anal. 8, 75-79. {MR44 #4902}

Keda, N.P. (1962): Chebyshev type quadratures for periodic functions (Russian), Vesci Akad. Navuk BSSR Ser. Fiz.-Tèhn. Navuk 1962, no.1, 19-23. {MR26 #2010}

Kozlovskiĭ, N. Ja. (1971): On the question of estimation of the remainder term of the Chebyshev formula (Russian), Dokl. Akad. Nauk BSSR 15, 965-967. {MR45#2919}

Krawtchouk, M. (1935): On an algebraic question in the moment problem (Ukrainian), J. Inst. Math. Acad. Sci. Ukraine 2, 87-92. {Zbl. 12, 294}

Krylov, V.I. (1957): On the proof of impossibility of constructing quadrature formulas with equal coefficients and more than nine nodes (Russian), Trudy Inst. Fiz. i Mat. Akad. Nauk BSSR no.2, 249-254. {Ref. Ž. (1958)#9269}

Krylov, V.I. (1958): Mechanical quadratures with equal coefficients for the integrals $\int_0^\infty e^{-x} f(x)dx$ and $\int_{-\infty}^\infty e^{-x^2} f(x)dx$ (Russian), Dokl. Akad. Nauk BSSR 2, 187-192. {MR22 #861}

Krylov, V.I. (1962): Approximate Calculation of Integrals, Transl. from Russian by A.H.Stroud, MacMillan, New York-London. [2nd ed. (Russian), Izdat, "Nauka", Moscow, 1967] {MR26 #2008, MR36 #1104}

Kuzmin, R.O. (1938): On the distribution of roots of polynomials connected with quadratures of Chebyshev (Russian), Izv. Akad. Nauk SSSR Ser. Mat. 2, 427-444. {Zbl. 19, 405}

Mayot, M. (1950): Sur la méthode d'intégration approchée de Tchebychef, C.R. Acad. Sci. Paris 230, 429-430. {MR11, 464}

Meir, A., and Sharma, A. (1967): A variation of the Tchebicheff quadrature problem, Illinois J. Math. 11, 535-546. {MR35 #7058}

Natanson, I.P. (1965): Constructive Function Theory, vol. III, Interpolation and Approximation Quadratures. Frederick Ungar Publ. Co., New York. {MR33 #4529c}

Nutfullin, Š.N., and Janovič, L.A. (1972): Čebyšev quadrature formulae with certain weight functions that depend on parameters (Russian), Vesci Akad. Navuk BSSR Ser. Fiz.-Mat. Navuk 1972, 24-30. {MR48 #9203}

Ossicini, A. (1966): Sulle formule di quadratura di Tschebyscheff, Pubbl. Ist. Naz. Appl. Calcolo, no.660, quad. 7, 43-59.

Ostrowski, A.M. (1959): On trends and problems in numerical approximation, in: "On Numerical Approximation" (R.E. Langer, ed.), pp.3-10. The University of Wisconsin Press, Madison. {MR20 #7381}

Posse, K.A. (1875): Sur les quadratures, Nouv. Ann. de Math. (2) 14, 49-62.

Rabinowitz, P., and Richter, N. (1970): Chebyshev-type integration rules of minimum norm, Math. Comp. 24, 831-845. {MR45 #7996}

Radau, R. (1880a): Sur les formules de quadrature à coefficients égaux, C.R. Acad. Sci. Paris 90, 520-523.

Radau, R. (1880b): Étude sur les formules d'approximation qui servent à calculer la valeur numérique d'une intégrale définie, J. Math. Pures Appl. (3)6, 283-336.

Riess, R.D., and Johnson, L.W. (1969): Estimating Gauss-Chebyshev quadrature errors, SIAM J. Numer. Anal. 6, 557-559. {MR41 #6398}

Rivlin, T.J. (1974): The Chebyshev Polynomials, Wiley, New York-London-Sydney-Toronto.

Rosati, F. (1968): Problemi di Gauss e Tchebychef relativi a formule di quadratura esatte per polinomi trigonometrici, Matematiche (Catania) 23, 31-49. {MR41 #1223}

Salkauskas, K. (1971): Existence of quadrature formulae with almost equal weights, Math. Comp. 25, 105-109. {MR44 #7750}

Salkauskas, K. (1973): Almost-interpolatory Chebyshev quadrature, Math. Comp. 27, 645-654 {MR49 #5658}

Salkauskas, K. (1975): On weight functions for Chebyshev quadrature, Numer. Math. 24, 13-18.

Salzer, H.E. (1947): Tables for facilitating the use of Chebyshev's quadrature formula, J. Math. and Phys. 26, 191-194. {MR9, 251}

Salzer, H.E. (1955): Equally weighted quadrature formulas over semi-infinite and infinite intervals, J. Math. and Phys. 34, 54-63. {MR16, 1055}

Salzer, H.E. (1957): Equally-weighted quadrature formulas for inversion integrals, Math. Tables Aids Comput. 11, 197-200. {MR19, 771}

Sard, A. (1963): Linear Approximation, Mathematical Surveys No.9, Amer. Math. Soc., Providence, R.I. {MR28 #1429}

Sonin, N. Ja. (1887): On the approximate evaluation of definite integrals and on the related integral functions (Russian), Warshawskia Univ. Izv. 1, 1-76.

Stetter, F. (1968a): Error bounds for Gauss-Chebyshev quadrature, Math. Comp. 22, 657-659. {MR37 #3763}

Stetter, F. (1968b): On a generalization of the midpoint rule, Math. Comp. 22, 661-663. {MR37 #2449}

Tureckiĭ, A.H. (1962): On the existence of Chebyshev quadrature formulas for an infinite interval (Russian), Vesci Akad. Nauk BSSR Ser. Fiz.-Tehn. Navuk 1962, no.2, 119-120. {Zbl.178, 516}

Ullman, J.L. (1962): Tchebycheff quadrature is possible on the infinite interval, Bull. Amer. Math. Soc. 68, 574-576. {MR26, #526}

Ullman, J.L. (1963): Tchebycheff quadrature on the infinite interval, Trans. Amer. Math. Soc. 107, 291-299. {MR26 #5330}

Ullman, J.L. (1966a): A class of weight functions for which Tchebycheff quadrature is possible, Bull. Amer. Math. Soc. 72, 1073-1075. {MR34#4766}

Ullman, J.L. (1966b): A class of weight functions that admit Tchebycheff quadrature, Michigan Math. J. 13, 417-423. {MR34 #5290}

Wilf, H.S. (1961): The possibility of Tschebycheff quadrature on infinite intervals, Proc. Nat. Acad. Sci. USA 47, 209-213. {MR23 #A2683}

Wilf, H.S. (1967): Advances in numerical quadrature, in: "Mathematical Methods for Digital Computers", vol. II (A. Ralston and H.S. Wilf, eds.), pp.133-144. Wiley, New York-London-Sydney. {MR35#2516}

Yanagiwara, H., and Shibata, K. (1974): A Chebyshev quadrature formula for twelve nodes, Bull. Fukuoka Univ. Ed. III 23, 45-51.

Yanagiwara, H., Fukutake, T., and Shibata, K. (1975): Chebyshev-type quadrature formulas for fourteen-seventeen nodes, Bull. Fukuoka Univ. Ed. III 24, 41-48.

Note added in proof. In addition to the references Ghizzetti and Ossicini [1970], Kozlovskiĭ [1971] in §7.1, mention should be made of the paper T. Popoviciu, "La simplicité du reste dans certaines formules de quadrature", Mathematica (Cluj) 6 (29) (1964), 157-184 {MR32 #4848}, in which the remainder is studied not only of the classical Chebyshev quadrature rule, but also of the Chebyshev-Laguerre and Chebyshev-Hermite formulas obtained by Salzer [1955].

Row Elimination for Solving Sparse Linear Systems and Least Squares Problems

W. Morven Gentleman

Introduction: What is row elimination?

Two of the most basic operations in modern computational linear algebra are the elementary stabilized transformation and the Givens transformation or plane rotation (the latter expressed either classically or in any of the newer, square root free forms [1,3]). Applied from the left to the rows of a matrix, these transformations are stable ways to combine a pair of rows to produce a new pair of rows, in which a specified element position in the original rows has been set to zero.

Repeated application of either of these transformations to a linear system, $Ax = b$, can obviously be used to transform the linear system to an equivalent one, $A'x = b'$, which is triangular and hence readily solved by backsubstitution. It is equally obvious that, given a linear least squares problem to minimize $||r||_2$, $r = y - Ax$, since Givens transformations are orthogonal and so do not change the L_2 norm, repeated application of such transformations can transform the problem to an equivalent one, to minimize $||r'||_2$, $r' = y' - A'x$, for which the matrix is triangular and hence for which the problem is also readily solved by backsubstitution. We will refer to either of these processes as row elimination.

So ingrained in the modern viewpoint is triangular decomposition or the cyclic QR reduction by rows or by columns, that it might seem at first that row elimination as just defined is merely a pedantic description of familiar processes such as Gaussian elimination with partial or complete pivoting, or the QR decomposition as conventionally described [1,7]. Examining the definition more closely, however, will reveal that while these familiar processes are included, so are many that are less familiar, for we have said nothing about the sequence of pairs of rows we will consider, nor the sequence of locations of zeros we will introduce. (Of course, we are unlikely to use a transformation that will destroy a zero we have deliberately created earlier). As an example of one of these less familiar processes that is row elimination but not triangular decomposition, there is a process described by Wilkinson [5, p. 17] for solving linear systems by sequentially reading the rows of the system, and using stabilized elementary transformations to eliminate each row with all the rows so far built into the triangle, the row finally being put into the triangle before the next row is read in. Since this process does not necessarily imply a single pivotal row associated with a column, it is clearly fundamentally different from triangular decomposition.

The numerical stability of row elimination based on stabilized elementary transformations was examined in a Ph.D. thesis by L. D. Rogers [6]. The backward error analysis bounds are not as satisfactory as those for triangular decomposition, because in addition to the factors present in the triangular decomposition bound, there is an additional growth factor arising strictly from the ordering of the eliminations, because the elementary elimination matrices may no longer commute. However, like the potential growth factor with partial pivoting, this growth factor is readily monitored and rarely is significant in practice, moreover, examples constructed to have a large growth factor invariably are accurately solved. General row elimination with elementary stabilized transformations appears as safe in practice as triangular decomposition with partial pivoting.

The analysis of numerical stability of row elimination based on Givens transformations is more satisfactory: the error bound [2] is essentially proportional to the number of non-commuting factors which the transformations applied must be written as, and is in any case therefore bounded by the number of transformations applied. Observed error growth is negligible - typically the error is not much larger than representation error for the final triangle.

There is an interesting asymmetry in viewpoint implicit in row elimination: whereas transformations from the left are central to the row elimination process for solving the given problem, transformations from the right are merely a change of basis for the coordinate system in which the solution is to be computed, and in many problems can be freely employed to make the matrix more amenable to work with. (Of course it is typically the case that if the solution is computed in a coordinate system other than originally specified, it must be transformed back to the original coordinate system before the computation is completed). We will return to discuss transformations from the right later.

Why is row elimination of interest?

Row elimination schemes are of interest in sparse matrix computations for several reasons. First and foremost, row elimination is a stable process of which conventional algorithms such as triangular decomposition are but special cases, and since zeros are as easily exploited by the general row elimination process as by the special case, there is some hope that pivoting schemes can be found that will produce elimination sequences that are cheaper, in terms of processor and fill, than more conventional approaches - and for which the savings will not be consumed by the pivoting scheme itself. Simple examples confirm the plausibility of this hope. For instance, matrices of the form of Figure 1 occur in the least squares analysis of factorial designs, and while row order is irrelevant, column order must be preserved. It is easy to see that if

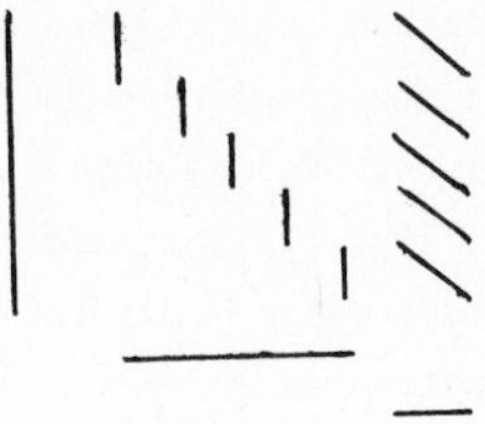

Figure 1

the problem is solved conventionally by Givens transformation QR decomposition, by rotating rows into a triangle sequentially, only the zeros on one side of the central group of columns can be exploited, whereas if each block of rows corresponding to a fixed nonzero location in the central group of columns is separately triangularized first, and then the triangles combined, essentially all zeros on both sides of the central group of columns will be exploited.

But there are other attractiveness to general row elimination. For example, the obvious flexibility in the order in which zeros are introduced can be an advantage itself. In some situations, for instance, a large proportion of the rows of the matrix are identical for a set of problems, and there are obvious advantages to transforming that part of the problem only once for all of the set, then using a copy of this to eliminate the remaining rows for each different problem.

Another attraction is the fact that with row elimination, the row is a logical entity which is treated at one time, moreso than different rows are treated simultaneously. This means that for computations on paged virtual memory machines, the row is a natural unit to store contiguously or at least in a single dynamic storage subpool in order to induce the principal of locality in data references and hence keep down paging activity.

For much the same reasons, when considering problems too large for main store, the fact that elements of a row are treated sequentially, and that rows interact, rather than arbitrary elements of the matrix interacting, makes row elimination potentially more attractive for use with backing store than some of the more conventional matrix factorizations.

The fact that general row elimination can involve large numbers of row eliminations that can be done simultaneously makes it of interest for parallel execution hardware, such as multi instruction stream multiprocessors or dataflow machines.

Finally, general row elimination is of interest because it has provoked an observation that raises serious questions about one of the sacred adages of the sparse matrix game, namely the adage that "accidental zeros" created by cancellation are so rare as to be not worth concerning oneself with. If we consider Figure 2a and Figure 2b, we see immediately that this is not true.

$$\begin{matrix} * & u & v & & \\ * & & & * & \\ * & & & & * \end{matrix}$$

Figure 2a

$$\begin{matrix} * & u & v & & \\ & \alpha u & \alpha v & * & \\ & \beta u & \beta v & & * \end{matrix}$$

Figure 2b

If the first row in Figure 2a is used to eliminate the other two, then the fill elements produced are as in Figure 2b, and if either of the other two rows is used to eliminate the remaining one, then it is algebraically obvious that not one but two zeros will be introduced - the underflow fault handler on your machine is apt to tell you the same thing in practise. This illustrates that serious local fill is not necessarily bad - if we recognize that it can be removed at a later step. Notice that either transformation from the left or from the right can be used to clean up the fill, the latter being a sort of adaptive version of the exhortation given to users to find a sparse way to express their problem!

The structure of row eliminatin codes

The approach we have chosen to use in the row elimination codes built so far is to triangularize whatever rows of the matrix are currently available, leaving a structure which could be used directly to solve the problem given so far, or to which additional rows could be added, to be further triangularized. Triangularization in the sense used here means that an ordering is established for the columns, and for each column there is at most one row with a nonzero in this column which does not also have a nonzero in some earlier column. Note that nothing is said about the rank or dimensionality of the problem: it is my observation that singularity of the matrix is a frequent occurrence in practical sparse matrix computations, for instance because of coding conventions for labelling columns which imply that certain columns cannot contain nonzeros. If there is a well determined subproblem, a sparse matrix code should solve it.

In choosing a sequence of eliminations, there are no necessary constraints except for the obvious one that in general we cannot complete all the transformations to introduce zeros into a particular column before we have completed all the transformations to introduce zeros into the earlier columns. Our current codes, however, are more restrictive: they procede sequentially by columns (depending on which version of the program, either in the given order, or in some other order chosen a priori, or choosing at each stage the best candidate from the columns that remain) and having selected a pivotal column, do all eliminations necessary to reduce the column to only one nonzero (other than nonzeros in rows with some nonzeros in earlier columns). Several ways exist for choosing the sequence of pair of rows to transform when cleaning out a column: the way we have experimented with most is to choose at each point the two shortest remaining rows, ties being resolved by taking the first. If the candidate rows are maintained in a heap ordered by number of nonzeros, this is very inexpensive.

We have investigated a number of different data structures. Interestingly, the data structures appropriate for solving problems never seem appropriate for building the matrix, leading each package to have a modular decomposition consisting of six routines:

1. INIT, an initialization routine.

2. ENTER, a routine to enter a nonzero into the initial structure. This routine is called for each nonzero in the matrix and right hand side (the right hand side is also sparse in many problems).

3. ORDER, a routine that takes the initial structure built by ENTER, and rebuilds it into the structure to be used in the actual triangularization and backsubstitution.

4. ELIM, a routine which performs the basic operation of elimination, given pointers to two rows in the structure.

5. TRI, a driver routine that accomplishes the triangularization by determining the sequence of row eliminations to be done, and repeatedly calling ELIM to do them.

6. BKSOLV, a routine which performs the final backsubstitution.

Experiments: Objectives and Results

Evaluating the performance of a complex piece of software such as a sparse matrix code is not something that can be done just by proving a few theorems about algorithms or by conducting a few experiments with artificial data. Such things provide useful insight, of course, but the performance of real codes are often strongly affected by implementation issues that theorems about algorithms ignore, and real world problems often have considerable structure which, even though not explicitly used by a code, may noticeably affect its performance. For this reason, we have conducted, and are continuing to conduct, considerable detailed experimentation into the real world behaviour of general row elimination schemes.

Here are the results of three typical experiments on two very different sparse linear least squares problems. Problem GARVIN is a 359 × 324 problem that arises from attempting to estimate molecular binding energy from measured heat of reaction in chemical experiments. It is very sparse, and remains sparse throughout the elimination process. Problem BIRCH is a 384 × 205 problem that arises from measuring the deviations from flatness of a disk. It starts very sparse, but exhibits considerable fill during elimination. Two methods of solution are used, QR decomposition using classical Givens transformations, and the augmented matrix (AM) approach using elementary stabilized transformations for the general row elimination. The AM approach means solving the least squares problem by solving the augmented system

$$\begin{bmatrix} I & A \\ A^T & 0 \end{bmatrix} \begin{bmatrix} r \\ x \end{bmatrix} = \begin{bmatrix} y \\ 0 \end{bmatrix}$$

Three different column orderings are used: the given ordering GO; static reordering SR, i.e. columns ordered in increasing numbers of initial nonzeros; and dynamic reordering DR, i.e. columns ordered during elimination by selecting from the remaining columns the one with fewest nonzeros. A simpler data structure is possible, and was used, when implementing GO and SR than is required for efficient implementation of DR. The shortest two rows in the column under consideration were picked to eliminate next.

Experiment 1: How do the QR and AM approaches compare for solving sparse linear least squares problems, when general row elimination is used for each?

Observations: Using data structures where the records for nonzeros are linked by explicit pointers, we observed the following

Store in Records

Problem	Ordering	Initial A.M.	Initial Q.R.	Maximum A.M.	Maximum Q.R.	Final A.M.	Final Q.R.
GARVIN	GO	1675	658	1930	6235	1910	5684
	SR	1675	658	1838	680	1717	582
	DR	1675	658	1829	677	1699	565
BIRCH	GO	2598	1107	9375	9865	7584	3211
	SR	2598	1107	11437	8133	9322	3453
	DR	2598	1107	8311	6919	6505	3139

CPU Time in Seconds

Problem	Ordering	Enter A.M.	Enter Q.R.	Order A.M.	Order Q.R.	Solve A.M.	Solve Q.R.
GARVIN	GO	2.635	.408	2.267	2.568	.205	16.407
	SR	2.560	1.104	.718	2.501	.497	.336
	DR	2.397	.466	7.020	2.334	.165	1.765
BIRCH	GO	2.040	.771	29.520	1.538	.287	36.570
	SR	1.954	1.269	25.157	1.522	.466	34.130
	DR	1.732	.743	18.980	1.548	.247	22.575

Conclusions: In standard Fortran, the data structure for orderings GO or SR takes three storage units per record for nonzeros, whereas the data structure for ordering DR takes four storage units. The reduction in records used by DR is not enough in either problem to outway this disadvantage. This is interesting for main store computations, but even moreso for problems requiring backing store, where GO or SR have substantial implementation advantages.

The QR approach takes less store, in both problems and with either static or dynamic reordering, than does the AM approach. The extra fill with the given ordering, especially for problem GARVIN, indicates the difficulties inherent in those statistical problems for which any column interchange is prohibited.

The AM approach takes less processor time in each case than the QR approach. In problem GARVIN the time to enter the matrix is a significant part of the total computation time, but with the greater fillin of problem BIRCH this is not true there. It is surprising that the relatively small decrease in number of nonzeros in problem BIRCH when DR is used rather than SR, would make such a large decrease in processor time.

No clearcut superiority of AM or QR over the other was demonstrated, but insufficient store is often a more serious problem than excessive processor requirement, as the latter merely increases the cost where the former may prevent solving the problem.

Experiment 2: What is the value of dropping elements computed as zero?

Observations: We can skip the transformation to zero an element and just drop the element if it is smaller than some test criterion (C1), or we can drop any element that is computed and is smaller than the test criterion, the former being cheaper and the latter being more effective. The test criterion could be complete cancellation (C2), but for Givens transformations the error analysis [2] can readily be used to show that the bound on the error will not be increased by more than a factor f if the test criterion (C3) is $6\varepsilon\sqrt{2(f^2-1)/M}\, s$, where ε is the fundamental rounding error of the machine, M is the length of a column, s is the L_2 norm of the column, and 6 would be replaced by some other constant for non classical versions of the transform. This criterion is conservative, and so looser one (C4) was obtained by omitting the factor 2/M within the square root. In the experiments, the factor f = 2 was used.

Store in Records

Problem	Criterion	GO				SR			
		init	max	final	regained	init	max	final	regained
GARVIN	(C1)	658	6717	6100	364	658	725	582	13
	(C2)	658	6709	6116	199	658	701	582	12
	(C3)	658	6235	5684	996	658	680	582	48
	(C4)	658	6175	5603	1018	658	680	582	51
BIRCH	(C1)	1107	10007	3227	26	1107	8107	3453	0
	(C2)	1107	10000	3227	0	1107	8107	3453	1
	(C3)	1107	9865	3211	188	1107	8133	3453	66
	(C4)	1107	9865	3211	201	1107	8132	3452	87

CPU Time in Seconds

Problem	Criterion	GO			SR		
		enter	order	solve	enter	order	solve
GARVIN	(C1)	2.574	.139	15.105	2.484	.435	.419
	(C2)	2.597	.167	22.088	2.548	.459	.374
	(C3)	2.568	.205	16.407	2.501	.497	.336
	(C4)	2.551	.199	16.500	2.579	.501	.352
BIRCH	(C1)	1.422	.212	30.156	1.606	.442	31.309
	(C2)	1.520	.257	36.627	1.515	.442	33.796
	(C3)	1.538	.287	36.570	1.522	.466	34.130
	(C4)	1.511	.274	36.472	1.519	.460	34.036

Not all dropping criteria were tried with DR, but with the recommended criterion 43 nonzeros were regained in problem GARVIN and 115 nonzeros in problem BIRCH.

Conclusions: Skipping transforms if the element to be zeroed is sufficiently small is a necessity, as transformations with elements that algebraically should be zero and are just rounding errors leads to underflows, overflows, etc. which can result in the program being thrown off the machine, or in inordinate time spent in the floating point fault handling routine. Checking every element computed reduced the maximum store required by up to 7%, but increased CPU time by up to 20%.

As might be expected, the greatest gains came when the matrix had to be processed in the given order - fill from having to use an inferior column ordering being partly offset by being able to eliminate some of this fill at a later stage. The gains are not as satisfactory as might be hoped, however, so perhaps a more elaborate scheme might be considered.

Note that dropping more elements may not decrease fill, as illustrated by SR with problem BIRCH. Dropping elements changes the nonzero count of a row, and so possibly the sequence of eliminations to be performed, and in this case the change was unlucky and the new sequence was not quite as good.

It is satisfying that the advantages accruing from dropping elemts are similar for (C3) and (C4). This suggests that we are correctly identifying nonzeros that algebraically should have been computed as zero, and that we are getting most of them. On the other hand, (C2) was much less effective, which is unfortunate as it is the only criterion applicable for stabilized elementary transformations.

Experiment 3: Compare the performance of a conventional data structure, with linked records forming a row, to that of a data structure using contiguous records per row, which avoids pointers but requires general dynamic storage allocation. (The contiguous record structure is appealing because by omitting the pointers it is smaller, it might be faster, and on a paged virtual memory computer it ensures locality of reference which a linked structure might easily lose).

Observations: Performance of the linked structure is straightfoward.

Performance of Linked Structure

Problem			Store in Records			CPU Time in Seconds		
			init	max	final	enter	order	solve
GARVIN	AM	GO	1675	1930	1910	2.635	.408	2.268
		SR	1675	1838	1717	2.560	1.104	.718
	QR	GO	658	6717	6100	2.574	.139	15.105
		SR	658	725	582	2.467	.432	.419
BIRCH	AM	GO	2598	9375	7584	2.040	.771	29.520
		SR	2598	11437	9322	1.954	1.269	25.157
	QR	GO	1107	10007	3227	1.422	.212	30.156
		SR	1107	8107	3453	1.606	.442	31.309

Contiguous records per row are often not convenient for entering the matrix, so pairs of records per nonzero are needed during the entry phase to get the extra links. The contiguous record structure used keeps the right hand side elements in the pool with the row, which the linked structure does not, so records are needed for these. Finally the quicksort used in ordering requires a stack which is an additional 2 $\log_2$ of the initial number of nonzeros. From these facts, and space requirements of the linked record structure, lower bounds can be computed on the pool size required with the contiguous record data structure.

Necessary Pool Size in Records

Problem			Enter	Solve	Equivalent Space to Linked Structure
GARVIN	AM	GO	4093	2613	3578
		SR	4093	2521	3440
	QR	GO	2054	7076	10434
		SR	2054	1084	1446
BIRCH	AM	GO	5991	9964	14651
		SR	5991	12026	17744
	QR	GO	3007	10391	15394
		SR	3007	8491	12544

In presenting the performance of the contiguous record structure we make a preliminary condensation by noting that, with the compacting storage manager used, CPU time is very accurately fitted by a constant plus a term proportional to the number of storage compactions. We present this linear function rather than the individual times.

Performance of Contiguous Structure

Problem			CPU Time in Seconds				pool/	pool/	pool/
			read	order	constant	linear	compactions	compactions	compactions
GARVIN	AM	GO	2.578	1.179	2.425	.0542	4093/17	5000/14	10000/3
		SR	2.572	1.893	.743	.0701	4093/3	5000/2	10000/0
	QR	GO	2.294	.387	18.006	.0443	7500/303	9000/120	10500/77
		SR	2.304	.649	.399	.0322	2060/2	2500/1	3000/1
BIRCH	AM	GO	1.705	2.068	35.754	.0929	10500/377	12750/135	15000/84
		SR	1.705	2.587	29.545	.0985	12500/267	15000/86	18000/50
	QR	GO	1.499	.593	32.703	.1559	10800/657	12750/192	15000/113
		SR	1.482	.833	34.009	.1049	8900/736	10500/229	12000/145

<u>Conclusions</u>: It is immediately evident that for problems, such as GARVIN, which exhibit little fill, the contiguous record structure may save no store because the pool required during the entry phase may exceed the equivalent space used by the whole solution with linked records. Even when space is saved, it is less than indicated here because the contiguous record code is bulkier.

Any space saving achieved is at a considerable cost in CPU time.

Counting the time spent in compaction, CPU times observed were up to four times higher. Even if we assumed a pool so large no compactions took place, so that only the constant matters, the CPU time with the contiguous records is slightly greater. Since a row linked structure code can readily be modified, with no loss of performance, to ensure the principal of locality by dividing the freelist into subpools of adjacent records, and associating groups of adjacent rows with specific subpools for getting and returning records, the contiguous record structure has no inherent advantage there.

In short, the experiment indicates that the contiguous record structure is in every way inferior to the conventional linked structure, and should be abandoned.

Open Questions

General row elimination is still a new and not well understood process, and many questions are as yet unresolved. Development of a code for use with backing store is in progress, the objectives being reasonable efficiency combined with portability across the major scientific machines. New selection rules are being studied: a particularly promising one for elementary stabilized transformations being to use the shortest row as the elimination in all cases for which it is stable, then the next shortest remaining row, and so forth. More elaborate ways to exploit removable fillin are being considered. We are still a long way from being ready to answer the basic question: how do these methods compare with more classical ones?

References

1. Gentleman, W. M., "Least Squares Computations by Givens Transformations Without Square Roots", JIMA, 12 (1973), pp. 329-336.

2. Gentleman, W. M., "Error Analysis of QR Decomposition by Givens Transformations", Linear Algebra and its Applications, to appear.

3. Hammarling, S., "A Note on Modifications to the Givens Plane Rotation", JIMA 13, (1974), No. 2, pp. 215-218.

4. Lawson, C. L. and Hanson, R. J., "Solving Least Squares Problems", Prentice-Hall, 1974.

5. Modern Computing Methods, Notes on Applied Science, No. 16, National Physical Laboratory, London 1961.

6. Rogers, L. D., "Optimal Paging Strategies and Stability Considerations for Solving Large Linear Systems", Ph.D. Thesis, University of Waterloo, 1973.

7. Wilkinson, J. H., "The Algebraic Eigenvalue Problem", London, Oxford University Press, 1965.

NONLINEAR LEAST SQUARES AND NONLINEARLY CONSTRAINED OPTIMIZATION

Philip E. Gill and Walter Murray

Introduction

The major part of this paper is concerned with the problem of minimizing a sum of squares ie

P1 $$\text{minimize } \{F(x) = f^T f\}, \quad x \varepsilon E^n,$$

where f is a $m \times 1$ vector of nonlinear functions. Such problems commonly arise, with $m >> n$, in choosing optimal parameters to fit a nonlinear model to experimental data.

There is a considerable body of literature on the numerical solution of P1, but there is still no universally satisfactory algorithm. The failure of current algorithms is due in part to particular problems being ill-posed, but failures and poor performance do occur on problems that have well-defined solutions.

The problem P1 is an unconstrained optimization problem and can be solved by methods which deal with general problems of this class. However, the special form of $F(x)$ and of its second derivatives suggests that a superior approach is possible. If g is the gradient vector of $F(x)$ and G its Hessian matrix, then we have

$$g = 2J^T f, \quad \text{where } J \text{ is the Jacobian of } f,$$

$$\text{and } G = 2J^T J + 2 \sum_{i=1}^{m} f_i G_i, \quad \text{where } G_i \text{ is the Hessian matrix of } f_i.$$

In practice, near the solution of P1, we have $||f|| \doteq 0$, which implies the approximation

$$G \doteq 2J^T J. \tag{1}$$

This approximation to G is the fundamental assumption of most algorithms for nonlinear least squares.

The need for a radical reappraisal of these algorithms is demonstrated by the relative performance of a general optimization algorithm and of a special least squares algorithm when applied to P1: on many problems the general algorithm works better! The paper summarizes research which was undertaken to explain this observation.

Newton-Type Algorithms for General Optimization

Optimization algorithms are nearly all of the following form. Given $x^{(k)}$, the kth estimate of the solution, a new estimate is obtained by first determining a direction of search $p^{(k)}$ and then a step-length $\alpha^{(k)}$ so that $x^{(k+1)} = x^{(k)} + \alpha^{(k)} p^{(k)}$. We shall not be concerned in this paper with the determination of $\alpha^{(k)}$ since a full description of this can be found in Gill and Murray (1974a). In fact, different algorithms use the same procedure for determining $\alpha^{(k)}$; what distinguishes them from each other is the definition and method of computation of $p^{(k)}$. Since for the remainder of this paper we shall only be concerned with a single iteration we have dropped the superfix k.

In Newton's method for general optimization p is determined by solving

$$Gp = -g. \tag{2}$$

In a practical algorithm this is inadequate when G is not positive-definite, but for the purpose of exposition this simple definition will suffice. A good Newton-type algorithm, such as that given by Gill and Murray (1974b), is robust, reliable and efficient. If, therefore, we could emulate the method in a satisfactory way without computing second derivatives, the resulting algorithm would also be robust, reliable and efficient for the problem P1.

The Gauss-Newton Method

The Gauss-Newton method is defined by the use of (1) in (2); hence the direction of search is determined by solving the following system of equations

$$J^T J p = -\tfrac{1}{2} g = -J^T f. \tag{3}$$

When this method works at all it is usually remarkably efficient; however, it is very unreliable and often fails. There have been many suggestions for modifying the algorithm so as to preserve its successful performances and mitigate the failures. Before we discuss these modifications it is important to define more precisely the implementation of the Gauss-Newton method so that failures shall not be due, as they often are, to deficiencies of the implementation.

An Effective Implementation of the Gauss-Newton Method

If $J^T J$ is singular then p is not uniquely defined by (3). Since the spectral condition number of $J^T J$ is the square of that of J then, *merely by forming $J^T J$*, we may have constructed a matrix which is singular with respect to the precision of the machine being used.

A better definition of p is the least squares solution of

$$Jp = -f. \tag{4}$$

By least squares solution we mean that p which minimizes $||Jp + f||_2$. If J has full rank p can be determined by the Businger-Golub algorithm, which performs the orthogonal factorization

$$J = Q^T \left[\begin{array}{c} R \\ \hdashline 0 \end{array}\right], \tag{5}$$

where Q is an $m \times m$ orthogonal matrix and R is an $n \times n$ upper triangular matrix. The vector p is then determined by back-substitution in the equations

$$Rp = -\bar{f},$$

where $\bar{f}$ is the $n \times 1$ vector consisting of the first elements of Qf. The algorithm can be implemented without retaining Q but we must store Q if we wish to do iterative refinement.

If J is rank-deficient the solution to (4) is not unique and a particular solution could be arbitrarily large. We therefore choose the solution of least length and this is unique. The vector can be determined by first factorizing J in the form

$$J = Q^T \underbrace{\left[\begin{array}{c:c} \overbrace{R}^{t} & 0 \\ \hdashline 0 & 0 \end{array}\right]}_{n} S, \tag{6}$$

where S is an $n \times n$ orthogonal matrix. (Pivoting is necessary but we have omitted the operations for simplicity). The search direction is now found by first determining the $t \times 1$ vector u by back-substitution in the equations

$$Ru = -\bar{f}, \tag{7}$$

where $\bar{f}$ is composed of the first t elements of Qf, and then p, given by

$$p = S^T \left[\begin{array}{c} u \\ \hdashline 0 \end{array}\right]. \tag{8}$$

In practice matrices are rarely exactly rank-deficient and, even if they were, error in the computation would obscure this fact. The rank must therefore be determined by a decision in the algorithm that all the elements in the block remaining to be reduced to upper triangular form are negligible in magnitude. The choice of tolerance, or threshold, for deciding what is negligible can have a substantial effect on the computation of p, as the following example illustrates.

<u>Example</u>

$$J = \begin{bmatrix} 1 & 0 \\ 0 & \varepsilon \end{bmatrix}, \qquad J^TJ = \begin{bmatrix} 1 & 0 \\ 0 & \varepsilon^2 \end{bmatrix}$$

$$g = 2\begin{bmatrix} f_1 \\ \varepsilon f_2 \end{bmatrix}, \qquad p = \begin{bmatrix} -f_1 \\ -\varepsilon^{-1} f_2 \end{bmatrix}$$

If f_1 and f_2 are similar in magnitude, we have

$$\frac{p^T g}{||p||\ ||g||} \doteq 0 \qquad \text{for } \varepsilon \text{ small.}$$

If ε is less than our tolerance, however, we have

$$J = \begin{bmatrix} 1 & 0 \\ 0 & 0 \end{bmatrix} \quad \text{and} \quad p = \begin{bmatrix} -f_1 \\ 0 \end{bmatrix},$$

ie the new p is almost orthogonal to the old p. Moreover, the new p will no longer be almost orthogonal to the gradient and in this respect it would seem better to set our threshold at a large value. In doing so, however, we may unnecessarily restrict the subspace in which p lies. (In the event that f_2 is very small the first p is adequate, of course).

Alternatives to the Gauss-Newton Algorithm

We return now to the alternatives should the Gauss-Newton algorithm fail despite careful implementation. We note that a necessary (but not sufficient) condition for its failure is that J is in effect rank-deficient. There are three possibilities: we can

i) introduce a means of biasing the search direction towards that of the steepest descent direction, $-g$;

ii) use a general minimization algorithm;

iii) include an estimate of $\sum f_i G_i$ in the approximation of G. (Since Newton-type algorithms are reliable for P1, the failure of Gauss-Newton must stem from neglecting this term.)

A Method which Biases the Search Direction: The Levenberg-Marquardt Algorithm

This is a popular algorithm for nonlinear least squares, being frequently recommended in survey papers. The attraction lies not so much in its efficiency but its persistence. Although it does not fail as frequently as the Gauss-Newton method it can have a very poor rate of convergence.

The search direction is defined as the solution of the equations

$$(J^T J + \lambda I)\, p = -J^T f, \tag{9}$$

the scalar λ being adjusted at each iteration according to the success of the previous search direction. It is also adjusted *within* an iteration until $x+p$ is a successful new estimate. Again, p should not be computed by forming the equations (9) but the equations will suffice as a definition since we will not be concerned in this section with computational error. One observation we can make from (9) is that for $\lambda > 0$ p is uniquely defined since $J^T J + \lambda I$ is non-singular.

In order to analyse the algorithm, we require additional notation. Let V be the $n\times(n-t)$ matrix whose columns span the null space of $J^T J$, such that $V^T V$ is I_{n-t}, the identity matrix of order $n-t$. Let W be the $n\times t$ matrix whose columns span the range of $J^T J$, such that $W^T W$ is I_t. We therefore have $JV=0$ and $W^T V=0$. Since any vector in E^n can be expressed as a linear combination of the columns of V and W, we have

$$p = p_1 + p_2, \text{ say,}$$

where $p_1 = Wu$, $p_2 = Vy$, u is $t\times1$ and y is $(n-t)\times1$. Substituting in (9) for p gives

$$J^T J Wu + \lambda Wu + \lambda Vy = - J^T f. \qquad (10)$$

Premultiplying by W^T gives

$$(W^T J^T J W + \lambda I)\, u = - W^T J^T f, \qquad (11)$$

which uniquely defines u. Premultiplying (10) by V^T gives

$$y = 0 .$$

Hence we have $p = Wu = p_1$ and $p \in \mathcal{R}(J^T J)$. The special character of this result appears more striking if we consider the search direction, $\bar{p}$, which a Newton-type method would provide. In that case we have

$$(J^T J + \varepsilon B)\, \bar{p} = - J^T f ,$$

where εB is $\sum_{i=1}^{m} f_i G_i$ with $||B|| = 1$ and ε a scalar.

Let

$$\bar{p} = \bar{p}_1 + \bar{p}_2 ,$$

where $\bar{p}_1$ is $W\bar{u}$ and $\bar{p}_2$ is $V\bar{y}$. Making the substitution as before we have

$$V^T B V \bar{y} = - V^T B W \bar{u} ,$$

so that in general $\bar{p}$ is not in the range of $J^T J$. Since $||\bar{y}||$ is not necessarily small compared with $||\bar{u}||$ the vectors p and $\bar{p}$ will not be similar.

Methods which Compute or Approximate the Second Derivatives of f.

Since Newton-type methods for the general unconstrained problem are reliable and efficient, we could abandon any hope of improvement and apply them directly to P1 whenever our Gauss-Newton algorithm fails. Three considerations should be borne in mind, however. First, in the methods which require analytical second derivatives we should need to supply a subroutine to compute G and this would entail forming $J^T J$. We have already explained why we should avoid this computation. Second, a subroutine that forms a finite-difference estimate to G would do so using a subroutine to evaluate g. It is better to approximate G by accumulating $J^T J$ and finite-difference estimates of $f_i G_i$ derived using the gradients of the f_i separately. The effects of both cancellation error and truncation error will then be less important when $||f||$ is small. Third, if $J^T J$ is not nearly rank-deficient we need not approximate or compute the term $\sum_{i=1}^{m} f_i G_i$.

Brown and Dennis (1970) describe an algorithm in which an approximation to $\sum_{i=1}^{m} f_i G_i$ is obtained using a quasi-Newton updating formula. They obtain their approximation by recurring approximations to each matrix G_i, and this requires considerable storage. We have, however, been given a verbal report that the term $\sum_{i=1}^{m} f_i G_i$ can be approximated directly. (In either event it is better not to use the identity matrix I as the initial approximation since we have shown that, if $J^T J$ is singular, the search directions lie wholly in the subspace spanned by the columns of $J^T J$. The directions in which we require information about the curvature of G lie in the null space of $J^T J$.) It is still an open question whether quasi-Newton formulae used in the manner described will effectively approximate $\sum_{i=1}^{m} f_i G_i$. The properties of quasi-Newton formulae often depend on the search direction being chosen in some specific manner. Moreover they rarely (except in the neighbourhood of the solution) provide a good element-by-element approximation. In their normal mode of use this may not be important (eg if we use $2G$ as our approximation to G or if we approximate G^{-1} by $G^{-1} + H$, where $Hg = 0$, then the search *direction* is unaffected). However, in the mode just described the effects may be critical to success.

In all the above situations the search direction p is defined as the solution to the equations

$$(J^T J + C)p = - J^T f, \qquad (12)$$

where C is a given symmetric matrix, provided $J^T J + C$ is sufficiently positive definite. If C is positive definite then so is $J^T J + C$ and a satisfactory computational scheme for computing p is as follows. First perform the Cholesky factorization $C = LL^T$, where L is a lower triangular matrix. The vector p can then be computed as the least squares solution of the equations

$$\begin{bmatrix} J \\ \hline L^T \end{bmatrix} p = \begin{bmatrix} -f \\ \hline 0 \end{bmatrix}. \qquad (13)$$

This avoids the formation of $J^T J$. However, if C is indefinite the situation is more complicated. If $J^T J + C$ is indefinite then a suitable course of action is to factorize C using the modified Cholesky algorithm described in Gill and Murray (1974b). This determines a lower triangular matrix L and a diagonal matrix E such that $C + E = LL^T$. The search direction can then be determined using the L in (13). The remaining case is where C is indefinite and $J^T J + C$ positive definite. In these circumstances we know of no way to derive p as the solution of a least squares problem. In the next section we describe an approach which circumvents the difficulty.

The New Algorithm

It has been shown that the algorithms discussed so far are unsatisfactory because the search directions can differ arbitrarily (even in the neighbourhood of the solution) from those determined by a known efficient method, or the method of computing p is subject to large numerical error. We shall now demonstrate that approximating $\sum_{i=1}^{m} f_i G_i$ by finite-differences is inefficient due to unnecessary evaluations of the gradients of f_i. In fact we shall show that, if $J^T J$ is singular, the approximation $G \doteq 2J^T J$ is not necessarily invalid.

Suppose V and W are the matrices defined earlier. If v and w are any $(n-t) \times 1$ and $t \times 1$ vectors, respectively, such that $v^T v = w^T w = 1$, then

$$GWw = 2J^T JWw + 2\sum_{i=1}^{m} f_i G_i Ww.$$

Provided $||f||$ is small enough, this reduces to

$$G\,Ww \doteq 2\,J^T J\,Ww.$$

Thus $2J^T J$ is a reasonable approximation to G in the space spanned by the columns of W. In this space it is unnecessary and therefore inefficient to estimate second derivatives of f. Consider, however, the vector

$$GVv = 2\,J^T J\,Vv + 2\sum_{i=1}^{m} f_i G_i Vv,$$

$$= 2\sum_{i=1}^{m} f_i G_i Vv.$$

It follows that for $||f|| > 0$, no matter how small $||f||$ is, it is dangerous to assume $2\,J^T J$ is an adequate approximation to G in the space spanned by the columns of V. In practice the distinction between V and W will not always be precise but this does not invalidate the general principle.

We proceed to describe a method for computing p assuming the matrix $\sum_{i=1}^{m} f_i G_i = \varepsilon B$, say, where $||B|| = 1$ and ε is a scalar, to be available. It will not be known initially whether $J^T J + \varepsilon B$ is positive definite so it is necessary to be able to determine this fact during the course of the computation. If $J^T J + \varepsilon B$ is positive definite then the search direction is given by

$$(J^T J + \varepsilon B)p = -J^T f. \tag{14}$$

Let the rank of $J^T J$ be t and let p be $p_1 + p_2$ where p_1 is Wu and p_2 is Vy, u and y being $t \times 1$ and $(n-t) \times 1$ vectors, respectively. Substituting for p in (14) gives

$$J^T JWu + \varepsilon\,B\,Wu + \varepsilon\,B\,Vy = -J^T f. \tag{15}$$

If we define $\bar{p}_1$ to be the least squares solution of least length of

$$J \bar{p}_1 = - f, \tag{16}$$

then it follows from premultiplying (15) by W^T that

$$\bar{p}_1 = p_1 + O(\varepsilon)$$

$$\text{and } \bar{u} = W^T \bar{p}_1 = u + O(\varepsilon).$$

We would compute $\bar{p}_1$ by first performing the factorization of J given in (6). This necessitates computing the matrix S and it can be shown that a suitable choice of W and V is given by

$$S = \begin{bmatrix} W^T \\ \hline V^T \end{bmatrix}.$$

Premultiplying (15) by V^T gives

$$\varepsilon V^T B V y = - \varepsilon V^T B W u .$$

Defining $\bar{y}$ as the solution of

$$\varepsilon V^T B V \bar{y} = - \varepsilon V^T B W \bar{u} , \tag{17}$$

we have $\bar{y} = y + O(\varepsilon)$ and $\bar{p}_2 = V\bar{y} = p_2 + O(\varepsilon)$.

If necessary we could obtain better approximations to p_1 and p_2 by substituting for the neglected terms. This does not involve any refactorizations since it simply perturbs the right-hand sides of (16) and (17). If $J^T J + \varepsilon B$ is positive definite then so is $V^T BV$ and $\bar{y}$ is determined by applying Cholesky's algorithm to $V^T BV$. If $V^T BV$ is not positive definite then neither is $J^T J + \varepsilon B$ and it is legitimate, indeed necessary, to alter our definition of p. An adequate search direction is obtained automatically by applying the modified Cholesky algorithm to $V^T BV$; this can,if necessary,also yield a direction of negative curvature.

The New Algorithm Without Second Derivatives

The algorithm described in the previous section evaluates B only to form the product BV. (If an $O(\varepsilon^2)$ approximation is required we shall also need the vector $B\bar{p}_1$). When a subroutine to evaluate B is not available it is possible to approximate εBV directly in far fewer gradient evaluations than would be required to approximate B, for which the work required is equivalent to the evaluation of n Jacobians. The scheme is as follows. Let the jth column of V be denoted by v_j. Compute the matrices A_i defined as

$$A_i = n \overset{\xleftarrow{\quad n-t \quad}\xrightarrow{}}{\begin{bmatrix} \nabla f_i(x+hv_1) & \nabla f_i(x+hv_2) & \cdots \\ & & \end{bmatrix}},$$

where h is the finite-difference step. It follows that

$$\sum_{i=1}^{m} f_i G_i V = \varepsilon BV = \frac{1}{h}\left(\sum_{i=1}^{m} f_i A_i - \bar{A} \right) + O(h) ,$$

where $\bar{A} = \left[J^T f \,\middle|\, J^T f \,\middle|\, \ldots . \right]$. Premultiplying by V^T gives

$$\varepsilon V^T BV = \frac{1}{h} C + O(h) ,$$

where $C = \sum_{i=1}^{m} f_i V^T A_i$. In general the matrix C will not be symmetric, but a symmetric approximation $\bar{C}$, say, can be obtained by setting $\bar{C} = \frac{1}{2}(C + C^T)$. It is possible to arrange the computation so that the matrices A_i, $i = 1,2, .., m$, are not actually formed, since this would require excessive storage. In fact the storage required is only $\frac{1}{2}(n-t)(n-t+1)$ locations.

The vector $\varepsilon V^T BW\bar{u}$ on the right-hand side of (17) can be approximated by the vector $\frac{1}{h} b$, where

$$b = \left\{ \sum_{i=1}^{m} f_i A_i^T \bar{p}_1 - \bar{A}^T \bar{p}_1 \right\} .$$

Having first determined $\bar{p}_1$, which does not require an estimate to B, we would then solve the equations

$$\bar{C} \hat{y} = - b$$

to obtain $\hat{y}$, an approximation to y. The search direction is then set to $\bar{p}_1 + \hat{p}_2$, where $\hat{p}_2 = V\hat{y}$. In practice the matrix J will not in general be exactly rank-deficient. Its approximate rank is determined by ignoring non-zero quantities below some preassigned threshold. The larger the threshold the better the condition of R can become and, it can be shown, the smaller the likelihood that $\bar{p}_1$ is almost orthogonal to the gradient. In the classical Gauss-Newton algorithm the larger the threshold is set, the smaller the dimension of the subspace to which the search direction is restricted. In the new algorithm this danger is removed and the condition of R can be controlled by suitable selection of the preassigned threshold.

The classical Gauss-Newton algorithm, if implemented correctly, sometimes works even if J is rank-deficient, so it is not always necessary to compute $\hat{y}$. We suggest, therefore, the following scheme. First compute $\bar{p}_1$ using a large tolerance in the factorization of J and use this as the direction of search. If satisfactory progress is not made then in the next iteration (or in the current one if no progress was made) augment $\bar{p}_1$ by $\hat{p}_2$, where $\hat{p}_2$ is computed in the manner just given. An alternative scheme is to compute a sequence of $(n-t)$ vectors $\hat{p}_2$, each of which requires one additional evaluation of the Jacobian. In the scheme just described the vector $\hat{p}_2$ lies in the space spanned by the columns of V. Using just one evaluation of the

Jacobian we could compute a component in the direction v_1. Similarly two evaluations will give a component in the space spanned by (v_1, v_2) etc. A satisfactory search direction may thus be found in fewer than (n-t) evaluations of the Jacobian.

A further advantage of the new algorithm over all others for nonlinear least squares is that it does not necessarily terminate if $||J^T f|| = 0$, $||f|| \neq 0$, since a direction of negative curvature can be determined should one exist.

So far we have assumed the existence of a subroutine to evaluate the Jacobian matrix. If such a routine does not exist the Jacobian matrix can be approximated either by finite-differences of f or by using a quasi-Newton updating scheme. Our own preference is to use finite-differences. It is not always appreciated that finite-difference methods whether for unconstrained optimization or nonlinear least squares are competitive with quasi-Newton methods provided the number of variables is small, say $n < 10$, and are not significantly worse in the range $10 < n < 20$. Most dense nonlinear least squares problems that we have encountered have no more than 20 variables. Moreover our experience is that the finite-difference algorithm is more reliable and the final approximation to the Jacobian, which is often required, is always better than that given by the quasi-Newton algorithm, which is sometimes in gross error. For larger problems the Jacobian usually has a significant number of zero or constant elements which can readily be exploited by a finite-difference algorithm. Our own experience has shown that in many problems some, but not all, of the analytical derivatives are known and these can also be readily incorporated in a finite-difference algorithm.

Given that we have an approximation to the Jacobian this can be used to obtain $\bar{p}_1$ from (16). The vector $\hat{p}_2$ can then be obtained by approximating the matrix $V^T BV$ directly just using evaluations of f. A total of $(n-t)(n-t+1)$ evaluations of f are required for this purpose and the right-hand side of (17) requires a further $n-t$. The number required to obtain $\hat{p}$ is, therefore, likely to be less than that to approximate the Jacobian.

Nonlinear Constrained Optimization

The remainder of this paper is concerned with the problem

P2 $$\min \{F(x)\} \qquad x \in E^n$$

subject to $$c_i(x) = 0 \quad i = 1, \ldots, m-1.$$

The need to be brief prevents us from giving the details of our proposals and from considering the implications of including inequality constraints.

An "equivalent" problem to P2 is the following:

$$\min \{\mathfrak{F}(x, \overset{*}{F}) = (F(x)-\overset{*}{F})^2 + c^T c\}, \ x \in E^n, \tag{18}$$

where $\overset{*}{F} = F(\overset{*}{x})$ and $\overset{*}{x}$ is the solution of P2. By "equivalent" problem we mean one which has a strong local minimum identical to a strong local minimum of P2. A key property of (18), which is rarely true with similarly proposed "equivalent" problems, is that the desired solution is the *global* minimum.

To our knowledge the first use of (18) was by Schmit and Fox (1965) who suggested approximating $\overset{*}{F}$ by $\hat{F}_1$, where $\hat{F}_1 > \overset{*}{F}$. They minimized $\mathfrak{F}(x, \hat{F}_1)$ and, having obtained the solution, they proceeded by subtracting some quantity from $\hat{F}_1$ to give $\hat{F}_2$, etc. Deciding how to adjust the estimate of $\overset{*}{F}$ presents an immediate difficulty and the technique they proposed would require many unconstrained minimizations. A second proposal was made by Morrison (1968) who suggested choosing $\hat{F}_1 < \overset{*}{F}$ and setting $\hat{F}_{i+1} = F(\bar{x}_i)$, where $\bar{x}^{(i)}$ is the minimum of $\mathfrak{F}(x, \hat{F}_i)$. It can be shown that if $\hat{F}_1 < \overset{*}{F}$ then so is $\hat{F}_i$, i=2,3,... and that $\hat{F}_i$ converges to $\overset{*}{F}$ at a linear rate. Again, many unconstrained minimizations may be necessary.

A third proposal, reportedly due to Wolfe, is to choose $\hat{F}_1 < \overset{*}{F}$ and to update using the formula

$$\hat{F}_{i+1} = \hat{F}_i + \frac{\mathfrak{F}(\bar{x}^{(i)}, \hat{F}_i)}{\left\{\mathfrak{F}(\bar{x}^{(i)}, \hat{F}_i) - \sum_{j=1}^{m-1} c_j(\bar{x}^{(i)})^2\right\}} . \tag{19}$$

It can be shown that this has a quadratic rate of convergence. Kowalik, Osborne and Ryan (1969) published some numerical results comparing the linear and quadratic estimates of $\overset{*}{F}$ and concluded that the quadratic ones were best. They were, however, somewhat fortunate in their examples, in that the quadratic formula did not overestimate $\overset{*}{F}$ as it could quite easily do. (The iteration (19) fails when $\hat{F}_i$ exceeds $\overset{*}{F}$ because the denominator in the second term then vanishes.) Their overall results compared reasonably well with those obtained using Penalty and Barrier function techniques; however, standards have improved somewhat since that time and the method does not appear to have taken root. Indeed, had we seen these results before undertaking our own experiments, we might well have been discouraged. For instance, one example required ten unconstrained minimizations to obtain a four decimal approximation to $\overset{*}{x}$, even starting from a good initial approximation to $\overset{*}{F}$. For this and similar approaches to be attractive the number of unconstrained minimizations must be small, say 5 or 6, even starting from a poor initial estimate. Moreover, the unconstrained problems must be solved efficiently with the latter minimizations requiring only a few function evaluations.

We have already noted the immediate shortcoming of (19), that the resulting estimate has no guarantee of being less than $\overset{*}{F}$. A second, less obvious, criticism is that as $\hat{F}_i \to \overset{*}{F}$ the formula becomes less reliable due to the combined effects of cancellation and of sensitivity to $\bar{x}^{(i)}$ being known inexactly. This second problem can be overcome by using the formula

$$\hat{F}_{i+1} = (1+\lambda^T\lambda)F(\bar{x}^{(i)}) - \hat{F}_i, \tag{20}$$

where λ is the vector of Lagrange multipliers. Since λ will not normally be known we can approximate it by $\bar{\lambda}$, where $\bar{\lambda}$ is the least squares solution to

$$A\,\bar{\lambda} = -g, \tag{21}$$

A being the Jacobian matrix of c and g the gradient of $F(x)$, both evaluated at the best approximation to $\overset{*}{x}$. If $\bar{x}^{(i)}$ is known exactly and this value is used, then (20) and (21) are theoretically equivalent to (19). The formula given in (20) provides the possibility of a better estimate to $\overset{*}{F}$ if better estimates of λ can be made. It can be shown that

$$\overset{*}{F} - F(\bar{x}^{(i)}) \doteq \frac{\lambda^T\lambda}{(\lambda^T\lambda+1)}(\overset{*}{F} - \hat{F}_i). \tag{22}$$

Since $\overset{*}{F}$ and $\hat{F}_i$ are fixed, $F(\bar{x}^{(i)})$ will be a closer approximation to $\overset{*}{F}$ if $\lambda^T\lambda$ is small. We can, of course, alter the length of λ by suitably scaling the constraints. However, if $\lambda^T\lambda$ were made very small, the method would simply become similar to the classical quadratic penalty function. A choice of $\lambda^T\lambda = 1/3$ would mean that even repeated use of the linear approximation to $\overset{*}{F}$ would reduce the error by 10^{-6} in ten minimizations.

After two or more minimizations it is possible by using extrapolation techniques to obtain better approximations to λ than that given by (21). Moreover, it is possible to obtain an even better approximation to $\overset{*}{F}$ than by using (20) and/or to estimate the error term so as to prevent $\hat{F}_{i+1}$ exceeding $\overset{*}{F}$. Implementing these improvements reduces the number of minimizations on the example mentioned earlier to four, even though a considerably worse initial estimate of $\overset{*}{F}$ was used. The accuracy at this point was also considerably improved, ten decimal places in x being correct.

Just as important as reducing the number of minimizations is efficient execution of the individual minimizations. If we define $f_i = c_i$, $i = 1,\ldots, m-1$ and $f_m = F-\hat{F}$, then $\mathfrak{F}(x, \hat{F})$ can be written

$$\mathfrak{F}(x,\hat{F}) = f^Tf.$$

The individual minimizations are therefore nonlinear least squares problems and we apply to $\mathcal{F}$ the notation developed in the earlier sections of the paper. Moreover, $\mathcal{F}$ tends to zero as $\hat{F} \to \overset{*}{F}$. In general the number of constraints active at the solution is less than $n-1$; hence the rank of J is normally less than n and J^TJ is singular for any $x \in E^n$. Moreover, if the Kuhn-Tucker conditions hold then (21) is satisfied at the solution and the last column of J^T is a linear combination of the first $m-1$ columns. Even if the Jacobian matrix of the constraint functions is of full rank, therefore, the rank of J at the solution is $m-1$. Clearly, for minimizing $\mathcal{F}$ it is essential to use a non-linear least squares algorithm of the type recommended in this paper. In fact, the algorithms proposed are ideally suited to this problem, since at the solution $\mathcal{F}=0$ and the rank of J will normally be known.

There are, of course, many other methods being developed for the nonlinearly constrained problem. We believe that the algorithm outlined here will prove to be important for the following reasons.

1. A good estimate of $\overset{*}{F}$ is often known in practical problems.

2. Most other algorithms depend heavily on the availability of accurate estimates of the individual Lagrange multipliers. These may be difficult, if not impossible, to estimate sensibly except close to the solution. If the Kuhn-Tucker conditions do not hold, such methods do not work.

3. Some methods require a feasible initial estimate of $\overset{*}{x}$, when solving inequality problems. If this is not provided it is usually determined by minimizing the function

$$\sum c_i^2(x),$$

the summation being made over the violated set. Obviously, in place of this we could equally well minimize

$$(F(x) - \hat{F})^2 + \sum c_i^2(x), \quad \hat{F} > \overset{*}{F}$$

which would then bias the feasible point found towards the solution.

4. Given a suitable nonlinear least squares subroutine the algorithm is relatively simple to implement.

5. The required solution is the global minimum. Usually with alternative techniques the required solution is only a local minimum and it could be that along many directions the transformed objective function tends to $-\infty$. (This objection to alternative techniques, which in our view is serious, could in many cases be overcome by replacing $F(x)$ by $(F(x)-\hat{F})^2$ in the transformed functions.)

6. The method can deal effectively with rank-deficiency in the Jacobian matrix of $c(x)$.

References

Brown, K.M. and Dennis, J.E. (1970) "New Computational Algorithms for Minimizing a Sum of Squares of Nonlinear Functions" Yale University Report.

Gill, P.E. and Murray, W. (1974a) "Safeguarded Steplength Algorithms for Optimization using Descent Methods" NPL Report NAC 31.

Gill, P.E. and Murray, W. (1974b) "Newton-type Methods for Unconstrained and Linearly Constrained Optimization" Math Prog 7, 311.

Kowalik, J., Osborne, M.R. and Ryan, D.M. (1969) "A New Method for Constrained Optimization Problems", Operations Research. 17, 973.

Morrison, D.D. (1968) "Optimization by Least Squares" SIAM J. Num. Anal. 5, 83.

Schmit, L.A. and Fox, R.L. (1965) "Advances in the Integrated Approach to Structural Synthesis", AIAA 6th Ann Struct. and Mat. Conf., Palm Springs.

Existence and Approximation of Weak Solutions of the Stefan Problem with Nonmonotone Nonlinearities

Joseph W. Jerome

Abstract

Consider the equation, in the distribution sense, for the temperature in a two-phase multidimensional Stefan problem

$$(1) \qquad \frac{\partial u}{\partial t} - \nabla.(k(u)\ \nabla u) + g(u) = f$$

on a space-time domain $D = (0,T) \times \Omega$ with specified initial and boundary conditions and enthalpy discontinuity across the free boundary. Here the conductivity coefficient k is a positive function with compact range, defined and continuous on R except at 0, and g is a Lipschitz body heating function, frequently encountered in welding problems, which is not assumed monotone. (We may take g such that $g(u)u \geq 0$).

Implicit two level time discretizations are employed in transformed versions of (1), giving a (finite) sequence of nonlinear elliptic boundary value problems (for each ∇t) which are solved by a Galerkin method. A subsequence of the step functions constructed on D is shown to converge weakly to a weak solution of the transformed equation. If, in addition, g is monotone, the entire sequence is strongly convergent to the unique solution.

Research supported by a grant from the Science Research Council, at Oxford University Computing Laboratory, 19 Parks Road, Oxford, OX1 3PL.

Introduction The mathematical model discussed in this paper arises from the two-phase Stefan problem in an arbitrary number of space dimensions. It is flexible enough to cover a number of situations in which such free boundary problems arise, such as the melting of ice and the welding of metals. In particular, we consider distribution diffusion equations for the temperature u of the form,

$$(1) \qquad \frac{\partial u}{\partial t} - \nabla.(k(u)\nabla u) + g(u) = f,$$

on a time-space domain $D = (0,T) \times \Omega$, where Ω is a bounded open set in R^N, $N \geq 1$, T is a fixed positive number, k is a positive function with compact range, defined and continuous on R^1 except at $\lambda = 0$, and g is the sum of two Lipschitz continuous functions g_1 and g_2 on R^1, with $g_1(\lambda)\lambda \geq 0$ and the Lipschitz constant of g_2 less than the smallest eigenvalue of $-\Delta$ on Ω with eigenfunctions vanishing on $\partial\Omega$; f is a given $L^\infty(D)$ function such that, for each $t \in (0,T)$, $f(., t) \in L^2(\Omega)$.

The discontinuity of the diffusion coefficient k at 0 corresponds to the change of phase at this temperature; we have chosen 0 for convenience. The presence of the function g may be interpreted as a body heating term; e.g., in welding problems, it arises from electrical resistivity and is termed a local joule heating effect. Also specified are a time independent function w, whose boundary values determine those of u, and an initial function u_o and a positive number b. When a classical solution, continuous and piecewise smooth, exists, then $b \cos(\nu, 1_t)$ represents the discontinuity of $k \frac{\partial u}{\partial \nu}$ normal to the bounding surface S of the time profiles D_1 and D_2 of the two phases. In this case, $u = o$ on S.

The problem (1) is a generalization of the classical Stefan problem,

$$(2) \qquad \frac{\partial u}{\partial t} - \nabla.(k(u)\nabla u) = f,$$

for which the notion of a weak solution was introduced by Oleinik [16]. She demonstrated that (2) is satisfied classically in D_1 and D_2, when $N = 1$, by the unique weak solution. These results were refined later by Douglas, Cannon and Hill [5] and by Friedman [8,9], the latter demonstrating the continuity of the solution and the fact that S is a continuous curve in this case.

Oleinik's weak or generalised solution formulation embeds boundary and initial conditions, thereby generalizing the usual notion of a distribution solution. Kamenomostskaja [11], via an explicit finite difference method, proved the existence of unique weak solutions of

(2) for general N and the mean square convergence of the step functions defined by the difference scheme. Both Oleinik and Kamenomostskaja transformed (2) by

$$(3) \qquad v = K(u) = \int_0^u k(\lambda)d\lambda,$$

giving an equation of the form,

$$(4) \qquad \frac{\partial H(v)}{\partial t} - \Delta V = f.$$

Here the enthalpy H is a discontinuous function at 0 and satisfies,

$$\text{(i)} \quad \frac{dH}{d\lambda} = \frac{1}{k(K^{-1}(\lambda))}, \quad \lambda \neq 0,$$

$$(5) \quad \text{(ii)} \quad H(0+) - H(0-) = b,$$

$$\text{(iii)} \quad H(0-) = 0 .$$

In the case where k is a piecewise constant function with values $0 < k_1$ for $\lambda < 0$, and $0 < k_2$ for $\lambda > 0$, then H is a piecewise linear function, with jump b at 0, satisfying $H'(\lambda) = \frac{1}{k_1}$ for $\lambda < 0$ and $H'(\lambda) = \frac{1}{k_2}$ for $\lambda > 0$. The method of Oleinik was to smooth H, yielding a sequence of quasi-linear parabolic equations, whose solutions were shown to converge uniformly, for N=1, to the solution of (2). The difference scheme of Kamenomostskaja was based on (4), treating H and v as a pair, with no direct reference to a free boundary. Friedman [8] later refined the existence results of Kamenomostskaja for the multidimensional Stefan problem by use of a smoothing method. In particular, he was able to treat time-dependent boundary conditions. Friedman also demonstrated additional regularity and stability of v: $\int |\nabla [v(x,t)]|^2$ is an essentially bounded function of t. This improved the result of [11], wherein v was known only to be a bounded measurable function for general boundary conditions.

By employing the transformation,

$$(6) \qquad v = B\,\Phi = H^{-1}(\Phi),$$

involving the inverse function H^{-1}, Brezis [3] reformulated (4) as

$$(7) \qquad \frac{\partial \Phi}{\partial t} - \Delta B\Phi = f.$$

Brezis assumed homogeneous boundary conditions, permitting multiplication by $E = (-\Delta)^{-1}$, which commutes with $\frac{\partial}{\partial t}$. Brezis thus obtained the standard form

$$(8) \qquad \frac{\partial (E\Phi)}{\partial t} + B\Phi = Ef$$

where E is a bounded, linear, self-adjoint, monotone operator on

$L^2(D)$ and B is strictly monotone, coercive and hemicontinuous. The existence of a unique weak solution of (8) is demonstrated in [3]. A slightly more general problem is discussed by Lions [12, p.196] whose proof, following [3], uses the constructive Faedo-Galerkin method.

A recent numerical analysis of the multidimensional Stefan problem has been carried out by Ciavaldini [4] and Meyer [13]. Both employ one-step time discretizations; Ciavaldini discretizes the weak formulation of (8) by a quadrature rule prior to employing implicit and explicit time approximations together with triangular finite elements. Stability is a consequence of the monotone formulation of the problem. Meyer, in the spirit of Oleinik and Friedman, smooths (4) prior to employing implicit time approximations together with finite difference approximations, defined via prolongation and restriction operators. Stability is assured by the maximum principle for quasilinear parabolic equations.

In this paper we apply implicit one-step time discretization to (1) as transformed by (3). Our results demonstrate both the existence of solutions for this more general problem as well as the L^2-convergence to a solution, unique if g is monotone, of a sequence of step functions, constructed from the solution of the unsmoothed nonlinear elliptic boundary value problems at each time-step. Novel existence proofs have been developed for these, depending on the existence and convergence of Galerkin approximations. Space limitations require us to present only the summary of our major results in section one. Complete proofs will appear elsewhere. We mention in closing that the paper of Atthey [2] served as our initial stimulation for problems involving the generalized formulation of (1). This and other topics are contained in the papers of [15] including a variational inequality formulation by Duvaut [6] and a summary of various approaches, including that of integral equations, by Tayler [18].

§1. Results of Existence and Convergence.

Let Ω be a bounded uniformly Lipschitz domain [14] in R^N, $N \geq 1$, and, for $T > 0$, let $D = (0,T) \times \Omega$. Suppose that (1) is given on D, with k, g and f as described in the introduction. If boundary conditions and initial data are specified in the form of functions w and u_0 in the Sobolev space $H^1(\Omega)$, let $W = K(w)$ and $U_0 = K(u_0)$ where K is given by (3). For $b > 0$ prescribed, let H be given by (5) and consider the transform of (1), effected by (3),

$$\frac{\partial H(v)}{\partial t} - \Delta v + G(v) = f \tag{1.1}$$

where $G(\lambda) = g(K^{-1}(\lambda))$, $\lambda \varepsilon R$. The initial and boundary conditions are specified precisely by

$$v(.\ ,0) = U_o,\quad v(.\ ,t) - W \varepsilon H_0^1\ (\Omega),\ \text{for each } t \varepsilon (0,T). \tag{1.2}$$

Here $H^1(\Omega)$ and $H_0^1(\Omega)$ have their usual meaning [1]. The sense in which we shall construct a solution of (1.1) is made precise by the following definition of weak solution.

Definition A bounded measurable function v on D is said to be a weak solution of the Stefan problem, if, for each $\phi \varepsilon C^\infty([0,T] \times \Omega)$ such that ϕ vanishes on $\{T\} \times \Omega$ and on $(0,T) \times \partial\Omega$, the identity

$$\int_D [H(v) \frac{\partial \phi}{\partial t} + v\, \Delta\phi - G(v)\phi + f\phi]\, dxdt - \int_{(0,T)\times\partial\Omega} W \frac{\partial \phi}{\partial \nu}\, d\sigma + \int_{\{0\}\times\Omega} H(U_o)\phi(x,o)\,dx = 0 \tag{1.3}$$

holds. This definition is due to Oleinik [16].

Theorem 1.1 Let D be a domain for which the divergence theorem holds and suppose that $U_o(x) \neq o$ a.e. in Ω and that W has $L^2\ (\partial\Omega)$ trace values $W(x) \neq o$ on $\partial\Omega$. Then, under the previously stated hypotheses on k, g and f, there exists a function $v \varepsilon L^\infty(0,T;\ H^1(\Omega))$ satisfying (1.3). v is unique if g is a monotone function.

The hypotheses on W and U_o ensure that H(W) is well defined on $\partial\Omega$ and $H(U_o)$ is well-defined on Ω. These conditions are necessitated by the fact that H, as defined by (5), remains undefined at 0. The hypotheses on W and U_o impose corresponding conditions on w and u_o. However, following Brezis [3] and Lions [12], we shall permit values of H(o) in (1.1) to lie in the interval [0,b]. With this interpretation, the solution of (1.3) is in fact a pair $[\Phi, v]$ related by (6).

It is necessary, to construct the approximate solutions of (1.3), to transform v by

$$v = V + W,\qquad V \varepsilon H_0^1(\Omega).$$

Since $-\Delta W$ is well-defined in the sense of distributions, we have from (1.1),

$$\frac{\partial H(V + W)}{\partial t} - \Delta V + G(V + W) = f + \Delta W. \tag{1.4}$$

The implicit difference scheme is based upon (1.4) and the equation

(1.5) $$V(\cdot\,,0) = U_o - W.$$

Suppose that a positive integer M is specified and set $\Delta t = T/M$. For m = 1, ... , M-1, consider the sequence of nonlinear elliptic boundary value problems given by

(1.6) $$[H(V_m + W) - H(V_{m-1} + W)]/\Delta t - \Delta V_m + G(V_m + W) = f_m + \Delta W,$$

(1.7) $$V_o = U_o - W.$$

Here f_m is the $L^2(\Omega)$ function $f(\cdot\,, m\Delta t)$.

The sense in which solutions of (1.6) and (1.7) are sought is made precise by the following variational formulation, which provides the basis for the Galerkin procedure. We seek, recursively for m = 1, ... , M-1, functions $V_m \varepsilon H_o^1(\Omega)$ satisfying, for all $\psi \varepsilon H_o^1(\Omega)$, the relation

(1.8) $$\int_\Omega \nabla V_m \cdot \nabla\psi + \left(\frac{1}{\Delta t}\right) \int_\Omega [H(V_m + W) + G(V_m + W)\Delta t]\,\psi = \left(\frac{1}{\Delta t}\right) \int_\Omega [f_m\,\Delta t + H(V_{m-1} + W)]\,\psi - \int_\Omega \nabla W.\,\nabla\psi.$$

For each m = 1, ... , M-1, (1.8) represents the weak formulation of (1.6).

<u>Theorem 1.2</u> For each m = 1, ... , M-1, there is a solution V_m of (1.8) in $H_o^1(\Omega)$, unique if g is a monotone function. Moreover, if $V^M = V^M(x,t)$ represents the step function on D defined by,

$$V^M(x,t) = V_m(x),\ x\varepsilon\Omega,\ m\Delta t \le t < (m+1)\Delta t,\ m = 0,1,\dots,M-1,$$

then $\{V^M\}_1^\infty$ is a subset of $L^\infty(0,T;\,H_o^1(\Omega))$ and there is a subsequence which is (strongly) convergent in $L^2(D)$ to a function V such that v = V + W is a solution of (1.3) provided the hypotheses of Theorem 1 hold. If g is a monotone function, $\{V^M\}_1^\infty$ is (strongly) convergent in $L^2(D)$ to a function V such that v = V + W is the unique solution of (1.3), i.e. the unique weak solution of the Stefan problem.

§2. Discussion

The method of proof of the existence of a solution of (1.8) involves the construction of an operator $T:H_o^1(\Omega) \to H^{-1}(\Omega)$, where $H^{-1}(\Omega)$ is the (topological) dual of $H_o^1(\Omega)$, of the form,

(2.1) $$\langle Ty, z\rangle = \int_\Omega \nabla y.\nabla z + \left(\frac{1}{\Delta t}\right) \int_\Omega [H(y+W) + G(y+W)\Delta t]z,$$

and the verification that T is surjective; the right hand side of (1.8) simply defines a member of $H^{-1}(\Omega)$.

Given $F \epsilon H^{-1}(\Omega)$, a sequence of smoothed problems

$$T_j y_j = F$$

is solved by a minimization technique introduced less generally in [7]; T_j is obtained from T by smoothing the enthalpy H. The sequence $\{y_j\}$ is bounded in $H^1(\Omega)$ and a subsequence converges appropriately to a solution y of Ty = F. The same smoothing technique shows that y may be approximated by Galerkin approximations of the unsmoothed problem, i.e., solutions z of,

$$P^T T P z = P^T F, \tag{2.2}$$

where P is a projection in $H^1_o(\Omega)$ onto a finite dimensional subspace. For finite element subspaces with local bases, the L^2 rate of convergence is not known without further hypotheses on y; eigenfunction subspaces, however, yield $O(n^{-1/N})$ order of L^2 convergence if g is a monotone function. This was pointed out for linear problems as long ago as 1970 [10]. It requires H^2 regularity of W and F however.

Stability, or the verification that $\{V^M\}_1^\infty \subset L^\infty(0, T; H^1_o(\Omega))$, is effected by special arguments using a generalized Gronwall inequality due to Raviart [17]. It is also shown that $\{V^M\}_1^\infty$ is relatively compact in $L^2(D)$ ensuring a convergent subsequence, say, with limit V. v = V + W is a solution of (1.3). We note in closing that the hypothesi that D is a domain for which the divergence theorem holds, may be eliminated if it is agreed to replace $\int_{(0,T)x\partial\Omega} W \frac{\partial\phi}{\partial\nu} d\sigma$ by $\int_D \nabla W.\nabla\phi + \int_D W\Delta\phi$.

References

1. S. Agmon, Lectures on Elliptic Boundary Value Problems, Van Nostrand, Princeton, New Jersey, 1965.

2. D.R. Atthey, A finite difference scheme for melting problems, J. Inst. Math.Appl. 13 (1974), 353-366.

3. H. Brezis, On some degenerate non linear parabolic equations, in Nonlinear Functional Analysis (F.E. Browder, editor) Part I, American Mathematical Society, Proceedings of Symposia in Pure Mathematics, Vol. 18, Providence, R.I., 1970, pp. 28-38.

4. J.F. Ciavaldini, Résolution Numerique d'un Probleme de Stefan a Deux Phases, Thesis, University of Rennes, France, 1972.

5. J. Douglas, J.R. Cannon and C.D. Hill, A multi-boundary Stefan problem and the disappearance of phases, J.Math.Mech. 17 (1967), 21-35.

6. G. Duvaut, The solution of a two-phase Stefan problem by a variational inequality,in Moving Boundary Problems in Heat Flow and Diffusion (J.R. Ockendon and W.R. Hodgkins, editors), Clarendon Press, Oxford, 1975, pp. 173-181.

7. S.D. Fisher and J.W. Jerome, Minimum Norm Extremals in Function Spaces with Applications to Classical and Modern Analysis, Springer Lecture Series in Mathematics, Heidelberg, 1975.

8. A. Friedman, The Stefan problem in several space variables, Trans. Amer. Math. Soc. 133 (1968), 51-87.

9. A Friedman, One dimensional Stefan problems with nonmonotonic free boundary, Trans. Amer. Math. Soc. 133 (1968), 89-114.

10. J.W. Jerome, On n-widths in Sobolev spaces and applications to elliptic boundary value problems, J. Math. Anal. Appl. 29 (1970), 201-215.

11. S.L. Kamenomostskaja, On the Stefan problem, Mat. Sb. 53 (1961), 489-514.

12. J.L. Lions, Quelques Méthodes de Resolution des Problèmes aux Limites non Linéaires, Dunod, Paris, 1969.

13. G.H. Meyer, Multidimensional Stefan Problems, SIAM J.Numer.Anal. 10 (1973), 522-538.

14. C.B. Morrey, Multiple Integrals in the Calculus of Variations, Springer-Verlag, New York, 1966.

15. J.R. Ockendon and W.R. Hodgkins (editors), Moving Boundary Problems in Heat Flow and Diffusion, Clarendon Press, Oxford, 1975.

16. O.A. Oleinik, A method of solution of the general Stefan problem, Soviet Math. Dokl. 1 (1960), 1350-1354.

17. P.A.Raviart, Sur l'approximation de certaines équations d'évolution linéaires et non linéaires, J.Math.Pures Appl. (9) 46 (1967), 11-183.

18. A.B. Tayler, The mathematical formulation of Stefan problems in Moving Boundary Problems, in Heat Flow and Diffusion, loc.cit., pp. 120-137.

On the Discovery and Description of Mathematical Programming Algorithms

Charles L. Lawson

I. INTRODUCTION

The primary purpose of this paper is to present some ideas on systematizing the description of mathematical programming algorithms. This approach will be illustrated by being used to describe a family of convex programming algorithms.

A high-level description will be given which characterizes the family of algorithms. Then two particular algorithms of the family will be described by refinement of the general high-level description. These two algorithms are respectively for the nonnegative least squares problem of minimizing $\|Ax-b\|$ subject to $x \geq 0$, and for the problem of minimizing $\|Ax\|$ subject to $x \geq 0$ and $\Sigma_i x_i = 1$.

This latter problem is treated by Wolfe [7] where an algorithm is given and some applications are mentioned. Besides the applications mentioned by Wolfe, this problem arises in the least squares estimation of Markov transition probabilities from time-series data [3].

II. METHODS OF DESCRIBING ALGORITHMS

Contemporary thinking regarding the design and description of computer programs (e.g., see [2] for survey papers and references) recognizes that there are many possible levels of description of a program. These range from possibly a one-sentence description of the function of a program to a fully detailed representation of the program in some programming language. In particular the notion of the top-down design of a program involves starting with a brief description of the program in very general terms and successively elaborating the specifications of the initially undefined operations. Various specific methodologies and syntactic forms have been developed to formalize the process of top-down program design and the associated record keeping.

I propose that a top-down mode of algorithm description and the systematic use of conventional syntactic forms could also be effectively used to improve the clarity of algorithm description in the published literature on mathematical programming.

Numerous algorithms in mathematical programming have the property of solving a sequence of equality-constrained optimization problems, eventually determining that one of these solutions is in fact the solution of the given problem. Such methods also typically use some updating method of solving the intermediate equality-constrained problems that economizes on computational operations by making use of information generated in solving the previous intermediate problems.

The mathematical properties of an algorithm that determine the sequence of intermediate problems to be solved are generally independent of the choice of updating method. A structured top-down description of an algorithm provides an effective way of preserving this independence in the description. This facilitates the comparison and classification of algorithms, the identification of the salient facts in convergence proofs, the derivation of new algorithms for specialized problems, and in general aids in human comprehension of an algorithm.

These ideas will be illustrated in the following sections.

III. A HIGH-LEVEL DESCRIPTION OF AN ALGORITHM FOR CONVEX PROGRAMMING

Let f be a real-valued convex C^1 function defined on R^n. Assume that f attains a minimum value in every linear flat, i.e., in every translated linear subspace. Let G be an $m \times n$ real matrix and h be a real m-vector. Let g_i^T denote the ith row vector of the matrix G.

Let $\mathcal{M} = \{1, \ldots, m\}$ and let $\mathcal{M}$ be partitioned into two disjoint subsets $\mathcal{E}$ and $\mathcal{I}$.

We will consider the following constrained minimization problem:

Problem A

$$\text{Minimize } f(x)$$

subject to

(1) $$g_i^T x = h_i \quad i \in \mathcal{E}$$

and

(2) $$g_i^T x \geq h_i \quad i \in \mathcal{I}$$

For any subset $\mathcal{J}$ of $\mathcal{M}$ let $G_{\mathcal{J}}$ denote the $m \times n$ matrix constructed by zeroing those rows of G not indexed in $\mathcal{J}$. Similarly let $h_{\mathcal{J}}$ denote an m-vector obtained by zeroing the components of h not indexed in $\mathcal{J}$.

With these notations Eq (1) and (2) can be expressed as

(3) $$G_{\mathcal{E}}\, x = h_{\mathcal{E}}$$

and

(4) $$G_{\mathcal{I}}\, x \geq h_{\mathcal{I}}$$

For any subset $\mathcal{J}$ of $\mathcal{M}$ define the linear flat

$$\mathcal{H}_{\mathcal{J}} = \{x : G_{\mathcal{J}}\, x = h_{\mathcal{J}}\} .$$

Note that if $\mathcal{J}$ and $\mathcal{K}$ are subsets of $\mathcal{M}$ with $\mathcal{J} \subset \mathcal{K}$ then $\mathcal{H}_{\mathcal{K}} \subset \mathcal{H}_{\mathcal{J}}$.

An over-bar denotes set complementation with respect to $\mathcal{M}$ and the symbol $\ominus$ denotes "and not". Thus

$$\mathcal{I} \ominus \mathcal{E} \equiv \mathcal{I} \cap \overline{\mathcal{E}}$$

A vector x will be called feasible if it satisfies Eq (3) and (4). The jth constraint will be called active, passive, or violated at a point x to indicate which of the conditions $g_j^T x = h_j$, $g_j^T x > h_j$, or $g_j^T x < h_j$ is true.

An algorithm, P1, for the solution of Problem A will be stated in very general terms. Algorithm P1 may be interpreted as a general statement of a number of different primal algorithms. We arrived at Algorithm P1 as a result of studing the common features of the equality and inequality constrained least squares algorithm of Stoer [6]*, the nonnegative least squares algorithm of Lawson and Hanson [4], and the algorithm of Wolfe [7] for nonnegative least squares with a unit sum constraint. All three of these algorithms can be stated as specializations of Algorithm P1. We expect that a number of other constrained minimization algorithms may also be interpreted as specializations of Algorithm P1.

It would be interesting and instructive as a means of classifying and understanding optimization algorithms to identify those that can be described as specializations of Algorithm P1. In the same vien, it would be interesting to identify other general algorithms which specialize to other families of particular algorithms.

Algorithm P1

```
INITIALIZE
Do until (converged)
          COMPUTE Z AND ALPHA
          If (hit a constraint) then
                    ADD CONSTRAINTS
          Else
                    KUHN-TUCKER TEST
                    If (.not. converged)
                              DROP A CONSTRAINT
                    End if
          End if
End do until
```

The names in capital letters in the statement of Algorithm P1 denote procedure calls. These five procedures are described as follows:

Procedure INITIALIZE

Let x be a feasible vector

$$\mathcal{J} := \{j : g_j^T x = h_j\}$$

converged := false

*Ref. [6] treats two problems, one with a positive-definite quadratic objective function and one whose objective function is a nonnegative-definite quadratic function plus a linear function. We refer here only to the algorithm for the former problem.

Procedure COMPUTE Z AND ALPHA

(5) Compute z to minimize f subject to $G_{\mathscr{J}}\ z = h_{\mathscr{J}}$

$\mathscr{A} :=$ $\{i : g_i^T z = h_i\}$ (Active set for z. Note $\mathscr{J} \subset \mathscr{A}$)

$\mathscr{V} :=$ $\{i : g_i^T z < h_i\}$ (Violated set for z)

hit a constraint := $\mathscr{V} \cup (\mathscr{A} \ominus \mathscr{J})$ is nonempty

If (hit a constraint)

$$\alpha := \min \left\{ \frac{g_i^T x - h_i}{g_i^T (x-z)} \ : \ i \in \mathscr{V} \cup (\mathscr{A} \ominus \mathscr{J}) \right\}$$

End if

Procedure ADD CONSTRAINTS

$x := x + \alpha(z-x)$

$\mathscr{K} := \{i : i \in \mathscr{V} \cup (\mathscr{A} \ominus \mathscr{J}) \text{ and } g_i^T x = h_i\}$

$\mathscr{J} := \mathscr{J} \cup \mathscr{K}$

Procedure KUHN-TUCKER TEST

$x := z$

$p :=$ gradient vector of f at x

Solve $[G_{\mathscr{J}}\ ]^T w = p$ for w

(This system is always consistent, but may have nonunique solutions. This nonuniqueness can substantially complicate an algorithm.)

If $w_i \geq 0$ for all $i \in \mathscr{J} \ominus \mathscr{E}$ then

  converged := .true.

Else

  $j :=$ an index $\in \mathscr{J} \ominus \mathscr{E}$ for which $w_j < 0$

End if

Procedure DROP A CONSTRAINT

$\mathscr{J} := \mathscr{J} \ominus \{j\}$

Although Algorithm P1 is stated in quite general terms, it is specific enough that a proof can easily be given that the algorithm converges, and that convergence occurs after only a finite number of repetitions of the Do until loop. The pattern of this proof is common in the mathematical programming literature. One shows that every time the first statement of the Do until loop is executed, the set $\mathcal{J}$ is different from all previous definitions of set $\mathcal{J}$. Each repetition of the Do until loop may be regarded as a "true" or "false" case depending on the outcome of the first If test. The false case can occur no more than m times in succession between true cases. On each true case, the value of f is strictly smaller than at the previous true case.

IV SPECIALIZATION OF ALGORITHM P1 FOR NONNEGATIVE LEAST SQUARES

As a first example of a specialization of Algorithm P1 we describe an algorithm for the following problem:

Problem NNLS

$$\text{Minimize } \|Ax-b\|$$

subject to

$$x \geq 0$$

The algorithm to be described is the same as Algorithm NNLS of Ref. [4]. I feel that the present structured presentation has some advantages relative to the description given in Ref. [4], although the description given in this short paper will not be as complete in all details as that given in Ref. [4]. For instance, it may be easier for readers to comprehend Algorithm NNLS when it is presented as a special case of Algorithm P1. Furthermore, this mode of description facilitates the derivation and programming of other algorithms related to NNLS as will be illustrated in Section V.

To establish the correspondence between Problem NNLS and the notation of Algorithm P1, take the objective function to be

$$f(x) = \|Ax-b\|^2 .$$

Then the gradient, p, at any point x is

$$p = A^T s$$

where

$$s = Ax-b .$$

The entities G, h, $\mathcal{I}$, and $\mathcal{E}$ of Eq. (3) - (4) will be defined by

(10)
$$[G : h] = [I : 0]$$

$$\mathcal{I} = \{1, \ldots, n\}$$

and

$$\mathcal{E} = \text{null} .$$

Next consider the operation required in COMPUTE Z AND ALPHA at Line (5). This is an equality constrained least squares problem. There are a number of reasonable methods for solving such a problem. For each method there are further choices available as to how to adapt the method to an "updating" mode for the economical solution of a sequence of such problems where each problem generally differs from the preceeding problem only by the addition or deletion of one index in the set $\mathcal{J}$. It is in these choices that one can make trade-offs involving execution speed, storage needed, numerical stability, appropriateness for cases of $k \gg n$ or $n \gg k$, appropriateness for sparse problems, etc.

We will assume that the data $[A : b]$ is given initially in a working array called $[W : y]$. The "updating" of the working array $[W : y]$ will be done by the application of orthogonal transformations to $[W : y]$ to triangularize the set of columns of W currently indexed in the set $\bar{\mathcal{J}}$. For brevity, we ignore details regarding permutation of columns of W and simply write

$$W = \begin{bmatrix} W_{11} & W_{12} \\ 0 & W_{21} \end{bmatrix}$$

where W_{11} denotes the triangularized columns indexed in $\bar{\mathcal{J}}$ and $\begin{bmatrix} W_{12} \\ W_{21} \end{bmatrix}$ denotes the columns indexed in $\mathcal{J}$.

The algorithm for NNLS consists of the same top-level algorithm as Algorithm P1 with the following definitions of the five procedures:

(NNLS)Procedure <u>INITIALIZE</u>

$x := 0$

$\mathcal{J} := \{1, \ldots, n\}$

converged := false

(NNLS)Procedure <u>COMPUTE Z AND ALPHA</u>

$$\begin{bmatrix} W_{11} & W_{12} & y_1 \\ 0 & W_{21} & y_2 \end{bmatrix}$$

Solve $[W_{11}, 0]\ z = y_1$ for z.

$\mathcal{A} := \{i : z_i = 0\}$

$\mathcal{V} := \{i : z_i < 0\}$

hit a constraint := $\mathcal{V} \cup (\mathcal{A} \ominus \mathcal{J})$ is nonempty

If (hit a constraint)

$$\alpha := \min \left\{ \frac{x_i}{x_i - z_i} : z_i \in \mathcal{V} \cup (\mathcal{A} \ominus \mathcal{J}) \right\}$$

(NNLS)Procedure <u>ADD CONSTRAINTS</u>

$x := x + \alpha (z - x)$

$\mathcal{K} := \{i : i \in \mathcal{V} \cup (\mathcal{A} \ominus \mathcal{J}) \text{ and } x_i = 0\}$

Do for all $i \in \mathcal{K}$

Remove column i from the triangularization

End do for

$\mathcal{J} := \mathcal{J} \cup \mathcal{K}$

(NNLS)Procedure KUHN-TUCKER TEST

x := z

$$\begin{bmatrix} W_{11} & W_{12} & y_1 \\ 0 & W_{22} & y_2 \end{bmatrix}$$

gradient of f at $x = p = -[0, W_{22}]^T y_2$

w := p

If $w_i \geq 0$ for all i then

converged := .true.

Else

j := an index for which $w_j < 0$

End if

(NNLS)Procedure DROP A CONSTRAINT

$\mathcal{J} := \mathcal{J} \ominus \{j\}$

Bring column j into the triangularization

V SPECIALIZATION OF ALGORITHM P1 FOR WOLFE'S CONSTRAINED LEAST SQUARES PROBLEM

As a second example of a specialization of Algorithm P1 consider the following problem treated in Wolfe [7]. Let A be a $k \times n$ matrix.

Problem W1

$$\text{Minimize } \|Ax\|$$

subject to

$$x \geq 0$$

and

$$\Sigma x_i = 1$$

Geometrically the problem is to find a convex combination of column vectors of A having least Euclidean length, i.e., lying closest to the origin in k-space.

The data is initially placed in a working array [W : y] as follows:

$$[W : y] = \begin{bmatrix} 1 \ldots 1 & 1 \\ & \vdots \\ A & 0 \end{bmatrix}$$

The algorithm can be taken to be the same as that for NNLS except for two of the procedures which are respecified as follows:

(W1) Procedure <u>INITIALIZE</u>

Let j be the index of the column of A having the minimal euclidean length.

Set $x_i := \begin{cases} 1 & \text{if } i = j \\ 0 & \text{if } i \neq j \end{cases}$

$\mathcal{J} := \{1, \ldots, n\} \ominus \{j\}$

Do Gaussian elimination pivoting in position (1, j)

converged := **false**

(W1) Procedure <u>ADD CONSTRAINTS</u>

Same as for NNLS except use Gaussian elimination if pivoting in the first row and use orthogonal transformation otherwise.

This algorithm for Problem W1 is different from the one given in Wolfe [7] although the Wolfe algorithm could also be described as a specialization of Algorithm P1, but with the details of the five procedures being different than we have given. Due to the use of orthogonal transformations [Ref. 4, 5], I expect that the algorithm given here would have better numerical accuracy, but require more execution time compared with Wolfe's algorithm.

VI. REMARKS AND WORK IN PROGRESS

1. The algorithm of Section V can be generalized further to handle the problem: minimize $\|Ax-b\|$ subject to $x \geq 0$ and $Cx = d$ where the constraints are nondegenerate.

2. The constraint $x \geq 0$ can be changed to constrain only specified components of x.

3. With algorithms for 1. and 2. and the introduction of slack variables, one can handle the problem: minimize $\|Ax-b\|$ subject to $Cx = d$ and $Ex \geq f$ where the constraints are nondegenerate.

4. The algorithm of Stoer [6] handles the problem of 3. and can be regarded as a specialization of Algorithm P1 but is different from the algorithm that would be derived as indicated in 3.

5. An algorithm permitting degenerate constraints can be devised that uses Algorithm P1 as a procedure.

VII. CONCLUSIONS

I feel that a top-down structured mode of description has merit for the comprehensible presentation of algorithms such as those of mathematical programming.

This approach also narrows the gap between algorithm description and computer programming if one has access to a programming language having the control structures used in these descriptions. I have found it a very satisfying experience to implement this family of algorithms using such a language: SFTRAN, a preprocessor for Structured Fortran due to John A. Flynn, Jet Propulsion Laboratory.

REFERENCES

1. Richard Bartels, Constrained Least Squares, Quadratic Programming, Complementary Pivot Programming and Duality, Proceedings of the 8th Annual Symposium on the Interface of Computer Science & Statistics,Health Science Computing Facility, Univ. of Calif., Los Angeles, Feb. 1975, pp. 267-271.
2. P.J. Denning, guest editor, ACM Computing Surveys, Special issue on programming, Vol 6, No. 4, (1974), pp. 209-319.
3. James K. Hightower, An Algorithm for Computing Restricted Least-Squares Estimates of Markov Transition Probabilities from Time-Series Data, Proceedings of the 8th Annual Symposium on the interface of Computer Science and Statistics, Health Science Computing Facility, Univ. of Calif., Los Los Angeles, Feb. 1975, pp. 238-241.
4. C.L. Lawson and R.J. Hanson, Solving Least Squares Problems, Prentice-Hall, Inc., (1974)
5. G.W. Stewart, Introduction to Matrix Computations, Academic Press, (1973)
6. Josef Stoer, On the Numerical Solution of Constrained Least-Squares Problems, SIAM J. Numer. Anal., Vol 8, No. 2, (1971), pp. 382-411.
7. Philip Wolfe, Algorithm for a Least-Distance Programming Problem, Mathematical Programming Study 1, (1974), pp. 190-205, North-Holland Publ. Co.

SOLUTION OF LINEAR COMPLEMENTARITY PROBLEMS BY LINEAR PROGRAMMING[1)]

O. L. Mangasarian

ABSTRACT

The linear complementarity problem is that of finding an $n \times 1$ vector z such that

$$Mz + q \geqq 0, \ z \geqq 0, \ z^T(Mz+q) = 0$$

where M is a given $n \times n$ real matrix and q is a given $n \times 1$ vector. In this paper the class of matrices M for which this problem is solvable by a single linear program is enlarged to include matrices other than those that are Z-matrices or those that have an inverse which is a Z-matrix. (A Z-matrix is real square matrix with nonpositive offdiagonal elements.) Included in this class are other matrices such as nonnegative matrices with a strictly dominant diagonal and matrices that are the sum of a Z-matrix having a nonnegative inverse and the tensor product of any two positive vectors in R^n.

1. INTRODUCTION

We consider the linear complementarity problem of finding a z in R^n such that

(1) $$Mz + q \geqq 0, \ z \geqq 0, \ z^T(Mz+q) = 0$$

where M is a given real $n \times n$ matrix and q is a given vector in R^n. Many problems of mathematical programming such as linear programming problems, quadratic programming problems and bimatrix games

[1)] Research supported by NSF grants GJ35292 and DCR74-20584

can be reduced to the above problem [4]. In addition some free boundary problems of fluid mechanics can be reduced to the solution of a linear complementarity problem [5,6,7]. The purpose of this paper is to extend the class of the matrices M for which the linear complementarity problem (1) can be solved by solving the single linear program

(2) minimize $p^T z$ subject to $Mz + q \geq 0,\ z \geq 0$

for an easily determined p in R^n. In [10] it was shown that for cases including those when M or its inverse is a Z-matrix, that is a real square matrix with nonpositive offdiagonal elements, the linear complementarity problem (1) can be solved by solving the linear program (2) for a certain p. In Section 2 of this paper we sharpen the main result of [10] by giving in Theorem 1 a characterization for the key condition which insures the solvability of the linear complementarity problem (1) by the linear program (2). Theorem 2 is a specific realization of Theorem 1 which has been given previously [10] in a slightly different form.

In Section 3 of the paper we extend further the class of linear complementarity problems solvable by a single linear program by considering an equivalent linear complementarity problem (7) with slack variables and by employing the results of Section 2. We obtain extensions which include cases such as when M is a nonnegative matrix with a strictly dominant diagonal or when M is the sum of a K-matrix, that is a Z-matrix having a nonnegative inverse, and the tensor product of any two positive vectors in R^n. A tabular summary of some of the linear complementarity problems solvable by a linear program is given at the end of the paper.

2. SOLUTION OF LINEAR COMPLEMENTARITY PROBLEMS BY LINEAR PROGRAMMING

In this section we shall characterize classes of matrices for which the linear complementarity problem (1) can be solved by solving the linear program (2). We begin by stating the dual to problem (2)

(3) maximize $-q^T y$ subject to $-M^T y + p \geq 0,\ \ y \geq 0$

and establishing the following key lemma [10] which insures that, under suitable conditions, any solution of the linear program (2) also solves the linear complementarity problem (1).

Lemma 1 If z solves the linear program (2) and if an optimal dual variable y satisfies

(4) $(I-M^T)y + p > 0$

where I is the identity matrix, then z solves the linear complementarity problem (1).

<u>Proof</u>: $y^T(Mz+q) + z^T(-M^Ty+p) = y^Tq + z^Tp = 0$.

Since $y \geqq 0$, $Mz + q \geqq 0$, $z \geqq 0$ and $-M^Ty + p \geqq 0$ it follows that

$$y_i(Mz+q)_i = 0,\ z_i(-M^Ty+p)_i = 0 \quad i = 1,\ldots,n$$

where subscripted quantities denote the ith element of a vector. But $y_i + (-M^Ty+p)_i > 0$, $i = 1,\ldots,n$, hence either $y_i > 0$ or $(-M^Ty+p)_i > 0$, $i = 1,\ldots,n$, and consequently $(Mz+q)_i = 0$ or $z_i = 0$, $i = 1,\ldots,n$. □

We give now a necessary and sufficient condition for the satisfaction of the key inequality (4) of Lemma 1.

<u>Theorem 1</u> Let the set $\{z \mid Mz+q \geqq 0, z \geqq 0\}$ be nonempty. A necessary and sufficient condition that the linear program (2) have a solution z with each optimal dual variable y satisfying (4) is that M, q and p satisfy

$$(5)\qquad \begin{aligned} MZ_1 &= Z_2 + qc^T \\ M^TX &\leqq pc^T \\ p^TZ_1 &> q^TX \\ X \geqq 0,\ c &\geqq 0,\ (Z_1,Z_2) \in Z \end{aligned}$$

and

$$(6)\qquad p = r + M^Ts,\ r \geqq 0,\ s \geqq 0$$

where X, Z_1, Z_2 are $n \times n$ matrices, c, r, s are vectors in R^n, and Z is the set of square matrices with nonpositive offdiagonal elements. If conditions (5) and (6) are satisfied then there exists at least one solution of the linear program (2), and each such solution solves the linear complementarity problem.

<u>Proof</u>: The existence of $(r,s) \geqq 0$ satisfying (6) is a necessary and sufficient condition for dual feasibility, which in turn is a necessary and sufficient condition that the feasible linear program (2) possess a solution. That each optimal dual variable y must satisfy (4), is equivalent to the system

$$Mz + q\zeta \geqq 0,\ z \geqq 0,\ -M^Ty + p\zeta \geqq 0,\ y \geqq 0,\ p^Tz + q^Ty \leqq 0,\ \zeta > 0$$
$$(M^T-I)_iy - p_i\zeta = 0$$

<u>not</u> having a solution (z,y,ζ) in R^{2n+1} for each $i = 1,\ldots,n$, where $(M^T-I)_i$ denotes the ith row of M^T-I . By Motzkin's transposition theorem [9] this in turn is equivalent to the existence of n-vectors

c and d, and $n \times n$ matrices X,Y,U,V and D, where D is diagonal, satisfying

$$M^TX + U - pc^T = 0$$

$$-MY + V - qc^T + (M-I)D = 0$$

$$q^TX + p^TY + d^T - p^TD = 0$$

$$(X,Y,U,V) \geqq 0,\ c \geqq 0,\ d > 0 \ .$$

By defining $Z_1 = D - Y$, $Z_2 = D - V$ these conditions become conditions (5).

The last statement of the theorem follows from Lemma 1. □

By taking $X = 0$, and $c = 0$ in (5) we obtain the following theorem which is equivalent to Theorem 1 of [10].

Theorem 2. Let the set $\{z|Mz+q\geqq 0,\ z\geqq 0\}$ be nonempty, and let M and p satisfy

$$MZ_1 = Z_2,\ p^TZ_1 > 0,\ (Z_1,Z_2) \in Z$$

$$p = r + M^Ts,\ r \geqq 0,\ s \geqq 0 \ .$$

Then the linear complementarity problem (1) has a solution which can be obtained by solving the linear program (2).

Useful special cases are obtained by setting $Z_1 = I$, $p = e$ and $Z_2 = I$, $p = M^Te$, where e is any positive vector and in particular it may be a vector of ones. In the first case we have that $M = Z_2 \in Z$, $p = e$, and in the second case that $M^{-1} = Z_1 \in Z$, $p = M^Te$. Other methods for solving (1) for Z-matrices and other related matrices are given in [1,2,3,6,11,12,13,14].

In order to enlarge further the class of matrices for which the linear complementarity problem can be solved by a linear program we consider a complementarity problem with slack variables which is equivalent to problem 1.

3. SOLUTION OF SLACK LINEAR COMPLEMENTARITY PROBLEMS BY LINEAR PROGRAMMING

We consider the following linear complementarity problem with a slack variable z_0 in R^m

$$(7) \qquad \begin{bmatrix} w \\ w_0 \end{bmatrix} = \begin{bmatrix} M & A \\ 0 & B \end{bmatrix} \begin{bmatrix} z \\ z_0 \end{bmatrix} + \begin{bmatrix} q \\ 0 \end{bmatrix} \geqq 0, \quad \begin{bmatrix} z \\ z_0 \end{bmatrix} \geqq 0, \quad z^Tw + z_0^Tw_0 = 0$$

where A is an $n \times m$ matrix and B is an $m \times m$ matrix.

Lemma 2. Let B be a strictly copositive or conegative matrix, that is $x^TMx \neq 0$ whenever $0 \leqq x \neq 0$. Then z solves the linear complementarity problem (1) if and only if $(z,z_0=0)$ solves the slack linear complementarity problem (7).

Proof: Obviously if z solves (1) then $(z,z_0=0)$ solves (7). If (z,z_0) solves (7) then since $0 = z_0^Tw_0 = z_0^TBz_0$, $z_0 \geqq 0$, and B is strictly copositive or conegative, then $z_0 = 0$ and z solves (1). □

By combining this lemma with Theorem 2 we can extend the class of matrices for which a linear program solves the linear complementarity problem.

Theorem 3. Let the set $\{z|Mz+q\geqq 0, z\geqq 0\}$ be nonempty, and suppose there exist Z_1,Z_2,Z_3,A,G,H,p and p_0 satisfying

(8) $MZ_1 = Z_2 + AG,\ MH \geqq AZ_3\ (Z_1,Z_2,Z_3) \in Z,\ (G,H) \geqq 0$

(9) $(p^T\ p_0^T)\begin{bmatrix} Z_1 & -H \\ -G & Z_3 \end{bmatrix} > 0 \qquad (p,p_0) \geqq 0$

where the dimensionalities of Z_1,Z_2,Z_3,A,G,H, and p_0 are respectively $n \times n$, $n \times n$, $m \times m$, $n \times m$, $m \times n$, $n \times m$ and $m \times 1$. Then the linear complementarity problem (1) has a solution which can be obtained by solving the linear program (2).

Proof: Set $B = I$ in problem (7) and apply Theorem 2 to it. In particular we have from (8) and (9) that

$$\begin{bmatrix} M & A \\ 0 & I \end{bmatrix}\begin{bmatrix} Z_1 & -H \\ -G & Z_3 \end{bmatrix} = \begin{bmatrix} Z_2 & -\bar{H} \\ -G & Z_3 \end{bmatrix}$$

$$(p^T\ \ p_0^T)\begin{bmatrix} Z_1 & -H \\ -G & Z_3 \end{bmatrix} > 0$$

where $\bar{H}$ is an $n \times m$ nonnegative matrix and

$$\begin{pmatrix} p \\ p_0 \end{pmatrix} = \begin{pmatrix} r \\ r_0 \end{pmatrix} \geqq 0 \qquad p_0 \in R^m,\ r_0 \in R^m .$$

Hence by Theorem 2 the slack linear complementarity problem (7) has a solution which can be obtained by solving the linear program

$$\text{minimize } p^Tz + p_0^Tz_0 \quad \text{subject to} \quad Mz + Az_0 + q \geqq 0,\ (z,z_0) \geqq 0 .$$

But since each solution of this linear program solves (7) it follows that $z_0 = 0$ at each solution of this linear program and hence we can set $z_0 = 0$ which reduces this linear program to (2). □

We observe that a sufficient condition for the inequality (9) to hold is that $\begin{bmatrix} Z_1 & -H \\ -G & Z_3 \end{bmatrix}^{-1} \geqq 0$. In fact this condition is also necessary for (9) to hold because the nonnegativity of the inverse $\bar{Z}^{-1}$ of a Z-matrix $\bar{Z}$ is equivalent to the existence of $p \geqq 0$ such that $p^T\bar{Z} > 0$ [8, Theorem 4,3]. Z-matrices with nonnegative inverses are called K-matrices [8] and sometimes M-matrices. The set of all K-matrices is denoted by K. By making use of these facts we can obtain the following consequence of Theorem 3.

<u>Theorem 4</u>. Let the set $\{z|Mz+q\geqq 0, z\geqq 0\}$ be nonempty, and let M satisfy

(10) $$M = (Z_2+AG)Z_1^{-1},\ MH \geqq AZ_3,\ (Z_1,Z_2,Z_3) \in Z,\ (G,H) \geqq 0$$

(11) $$Z_1^{-1} \geqq 0,\ (Z_3-GZ_1^{-1}H)^{-1} \geqq 0 \ .$$

Then there exists $(p,p_0) \in R^{n+m}$ satisfying (9) and the linear complementarity problem (1) has a solution which can be obtained by solving the linear program (2).

<u>Proof</u>: We will show that the conditions of Theorem 3 hold and hence (1) has a solution and is solvable by the linear program (2). We have that

$$\begin{bmatrix} Z_1 & -H \\ -G & Z_3 \end{bmatrix}^{-1} = \begin{bmatrix} Z_1^{-1}(I+HC^{-1}GZ_1^{-1}) & Z_1^{-1}HC^{-1} \\ C^{-1}GZ_1^{-1} & C^{-1} \end{bmatrix}$$

where $C = Z_3 - GZ_1^{-1}H$. It follows from $Z_1^{-1} \geqq 0$, $C^{-1} \geqq 0$, $H \geqq 0$ and $G \geqq 0$ that $\begin{bmatrix} Z_1 & -H \\ -G & Z_3 \end{bmatrix}^{-1} \geqq 0$ and that $(p^T\ p_0^T) = e^T \begin{bmatrix} Z_1 & -H \\ -G & Z_3 \end{bmatrix}^{-1} \geqq 0$, where e is any positive vector in R^{n+m}, satisfies (9). Conditions (8) follows from (10). □

By setting $Z_1 = I$ in the above theorem and defining $Z_4 = Z_3 - GH$ we obtain the following theorem.

<u>Theorem 5</u>. Let the set $\{z|Mz+q\geqq 0, z\geqq 0\}$ be nonempty, and let M satisfy

(12) $$M = Z_2 + AG,\ MH \geqq AZ_3 \qquad GH = Z_3 - Z_4 \geqq 0$$

(13) $$(Z_2,Z_3) \in Z,\ Z_4 \in K, \qquad (G,H) \geqq 0$$

then there exists $(p,p_0) \in R^{n+m}$ satisfying (9) with $Z_1 = I$ and the linear complementarity problem (1) has a solution which can be obtained by solving the linear program (2).

Note that since $Z_3 = Z_4 + GH \geqq Z_4$ and Z_4 is a K-matrix, it follows by Theorem 4,6 of [8] that Z_3 is also a K-matrix.

We conclude by giving some specific realizations of Theorem 5.

Theorem 6. Let the set $\{z \mid Mz+q \geqq 0, z \geqq 0\}$ be nonempty and let M satisfy any of the conditions below. Then the linear complementarity problem (1) has a solution which can be obtained by solving the linear problem (2) with the p indicated below:

(a) $M = Z_2 + ab^T$, $Z_2 \in K$, $a \in R^n$, $b \in R^n$, $0 \neq a \geqq 0$, $b > 0$, $p = b$.

(b) $M = Z_2 + A(Z_3 - Z_4)$, $(Z_2, Z_3) \in Z$, $Z_4 \in K$, $Z_3 \geqq Z_4$, $Z_2 \geqq AZ_4$, $p_0^T Z_4 > 0$, $p_0 > 0$, $p^T = p_0^T(Z_3 - \frac{1}{2}Z_4)$

(c) $M = 2Z_2 - Z_4$, $Z_2 \in Z$, $Z_4 \in K$, $Z_2 \geqq Z_4$, $p_0^T Z_4 > 0$, $p_0 > 0$, $p^T = p_0^T M$

(d) $M \geqq 0$, $M_{jj} > \sum_{\substack{i=1 \\ i \neq j}}^{n} M_{ij}$, $j = 1,\ldots,n$, $p^T = e^T M$, $e^T = (1,\ldots,1) \in R^n$.

(e) $M \geqq 0$, $M_{ii} > \sum_{\substack{j=1 \\ j \neq i}}^{n} M_{ij}$, $i = 1,\ldots,n$, $p^T = p_0^T M$ where $p_0^T Z_4 > 0$,

$p_0 > 0$ and $Z_4 = -M + 2(\text{diagonal of } M)$.

Proof: (a) Since $Z_2 \in K$, there exists an h in R^n, $h > 0$, such that $Z_2 h > 0$ [8, Theorem 4,3]. Set in Theorem 5 above: $A = a$, $G = b^T$, $H = h$, $Z_4 = \frac{1}{2}\min_j \frac{(Z_2 h)_j}{a_j > 0} > 0$ and $Z_3 = b^T h + Z_4$. Note that here Z_3 and Z_4 are real numbers. We now have that

$$MH - AZ_3 = (Z_2 + ab^T)h - a(b^T h + Z_4) = Z_2 h - aZ_4 > 0 .$$

To satify inequality (9), which in this case is $p^T > p_0 b^T$ and $p_0^T Z_3 > p^T h$, set $p = b$ and take p_0 satisfying $1 > p_0 > \frac{b^T h}{Z_3}$. Application of Theorem 5 gives the desired result.

(b) Set in Theorem 5, $H = I$. Conditions (12) and (13) are satisfied. Inequality (9) requires that

$$(p^T \; p_0^T) \begin{bmatrix} I & -I \\ -(Z_3 - Z_4) & Z_3 \end{bmatrix} > 0 .$$

That is we require that

$$p_0^T Z_3 > p^T > p_0^T (Z_3 - Z_4) .$$

Now we have that $p_0^T Z_4 > 0$, $p_0 > 0$, $Z_3 - Z_4 \geqq 0$ and hence

$$p_0^T Z_3 > p_0^T (Z_3 - Z_4) \geqq 0 \quad .$$

But since $p^T = p_0^T(Z_3 - \frac{1}{2}Z_4) > 0$ is the average of the first two terms in the above inequalities, it follows that the desired inequality

$$p_0^T Z_3 > p^T = p_0^T(Z_3 - \tfrac{1}{2}Z_4) > p_0^T(Z_3 - Z_4)$$

holds.

(c) Set $A = I$ and $Z_2 = Z_3$ in part (b) of this theorem, and take $p^T = p_0^T M$ instead of $p^T = \frac{1}{2}p_0^T M$ since this change does affect the solution of (2).

(d) Take part (c) of this theorem and set

$$(Z_2)_{ij} = 0,\ i \neq j,\ (Z_2)_{jj} = M_{jj},\ i,j = 1,\ldots,n$$

$$(Z_4)_{ij} = -M_{ij}, i\neq j,\ (Z_4)_{jj} = M_{jj},\ i,j = 1,\ldots,n \ .$$

The matrix Z_4 is a K-matrix because, for $j = 1,\ldots,n$, $M_{jj} - \sum_{\substack{i=1 \\ i\neq j}}^{n} M_{ij} > 0$.

Hence $p_0^T = e^T = (1,\ldots,1) \in R^n$ satisfies $p_0^T Z_4 > 0$. Take $p^T = p_0^T M = e^T M$.

(e) We again apply part (c) of this theorem and define

$$(Z_2)_{ij} = 0,\ i \neq j,\ (Z_2)_{ii} = M_{ii},\ i,j = 1,\ldots,n \ .$$

The matrix Z_4 is a K-matrix because, for $i = 1,\ldots,n$, $M_{ii} - \sum_{\substack{j=1 \\ j\neq i}}^{n} M_{ij} > 0$.

Hence there exists a $p_0 > 0$ such that $p_0^T Z_4 > 0$. Take $p^T = p_0^T M$. □

Note that in both cases (d) and (e) above, that is when M is a nonnegative strictly diagonally dominant matrix, $p^T = p_0^T M$, where $p_0 > 0$ is determined from the matrix Z_4 obtained from M by reversing the sign of the offdiagonal elements of M and requiring that $p_0^T Z_4 > 0$.

We close with a summary given in Table 1 below which gives the required assumptions on M and the corresponding vector p used in the linear program to obtain a solution of the linear complementarity problem. It is hoped that further research will substantially enlarge this table.

TABLE 1

Linear Complementarity Problems Solvable by Linear Programming

Matrix M of (1)	Conditions on M	Vector p of (2)	Conditions on p
$M = Z_2 Z_1^{-1}$	$Z_1 \in K,\ Z_2 \in Z$	p	$p \geqq 0,\ p^T Z_1 > 0$
$M = Z_2 Z_1^{-1}$	$Z_1 \in Z,\ Z_2 \in K$	$p = M^T s$	$s \geqq 0,\ s^T Z_2 > 0$
M	$M \in Z$	p	$p > 0$
M	$M^{-1} \in Z$	$p = M^T e$	$e > 0$
$M = Z_2 + ab^T$	$Z_2 \in K$ $0 \neq a \geqq 0,\ b > 0$	$p = b$	
$M = 2Z_2 - Z_4$	$Z_2 \in Z,\ Z_4 \in K$ $Z_2 \geqq Z_4$	$p = M^T p_0$	$p_0 > 0,\ p_0^T Z_4 > 0$
M	$M \geqq 0$ $M_{jj} > \sum_{i=1, i \neq j}^{n} M_{ij},\ j=1,\ldots,n$	$p = M^T e$	$e^T = (1,\ldots,1)$
M	$M \geqq 0$ $M_{ii} > \sum_{j=1, j \neq i}^{n} M_{ij},\ i=1,\ldots,n$	$p = M^T p_0$	$p_0 > 0,\ p_0^T Z_4 > 0$ $Z_4 = -M + 2\text{diag } M$

REFERENCES

1. R. Chandrasekaran, "A special case of the complementary pivot problem," Opsearch 7, 1970, 263-268.

2. R. W. Cottle & R. S. Sacher, "On the solution of large, structured linear complementarity problems: I," Technical Report 73-4, Department of Operations Research, Stanford University, 1973.

3. R. W. Cottle, G. H. Golub & R. S. Sacher, "On the solution of large, structured linear complementarity problems: III," Technical Report 74-439, Computer Science Department, Stanford University, 1974.

4. R. W. Cottle & G. B. Dantzig, "Complementary pivot theory of mathematical programming," Linear Algebra and Appl. 1, 1968, 103-125.

5. C. W. Cryer, "The method of Christopherson for solving free boundary problems for infinite journal bearings by means of finite differences," Math. Computation 25, 1971, 435-443.

6. C. W. Cryer, "The solution of a quadratic programming problem using systematic overrelaxation," SIAM J. Control 9, 1971, 385-392.

7. C. W. Cryer, "Free boundary problems," forthcoming monograph.

8. M. Fiedler & V. Pták, "On matrices with nonpositive off-diagonal elements and positive principal minors," Czech. J. Math. 12, 1962, 382-400.

9. O. L. Mangasarian, "Nonlinear programming," McGraw-Hill, New York, 1969.

10. O. L. Mangasarian, "Linear complementarity problems solvable by a single linear program," University of Wisconsin Computer Sciences Technical Report #237, January 1975.

11. R. Sacher, "On the solution of large, structured linear complementarity problems: II," Technical Report 73-5, Department of Operations Research, Stanford University, 1973.

12. R. Saigal, "A note on a special linear complementarity problem," Opsearch 7, 1970, 175-183.

13. R. Saigal, "Lemke's algorithm and a special linear complementarity problem," Opsearch 8, 1971, 201-208.

14. A. Tamir: "Minimality and complementarity properties associated with Z-functions and M-functions," Math. Prog. 7, 1974, 17-31.

SPARSE IN-CORE LINEAR PROGRAMMING

J.K. Reid

Summary Linear programming in core using a variant of the Bartels-Golub decomposition of the basis matrix will be considered. This variant is particularly well-adapted to sparsity preservation, being capable of revising the factorisation without any fill-in whenever this is possible by permutations alone. In addition strategies for column pivoting in the simplex method itself will be discussed and in particular it will be shown that the "steepest edge" algorithm is practical. This algorithm has long been known to give good results in respect of number of iterations, but has been thought to be impractical.

Test results on genuine problems with hundreds of rows and thousands of columns will be reported. These tests include comparisons with other methods.

1. Introduction

We will consider the solution of the standard linear programming problem of minimizing

$$c^T x \tag{1.1}$$

subject to contsraints

$$Ax = b \tag{1.2}$$

and

$$x \geq 0 \tag{1.3}$$

where A is an mxn matrix of rank m with m<n. The set of all feasible vectors x (those satisfying constraints (1.2) and (1.3)) forms a simplex in n-dimensional Euclidean space E^n and (unless it is null) a vertex of this simplex is a point at which the objective function (1.1) is minimized. The simplex method consists of a structured search of the vertices at each of which there are (m-n) variables having the value zero (the out-of-basis variables) and the remainder (the in-basis variables) are determined by equation (1.2). The search proceeds from vertex to adjacent vertex along an edge on which just one of the vertex's out-of-basis variables is non-zero.

Unless the problem is quite small it is usual to hold the basis B (set of columns of A that correspond to the in-basis variables) in a product form that allows easy solution of sets of equations

$$By = a \tag{1.4}$$

and

$$B^T z = d. \tag{1.5}$$

It is always necessary to solve (1.4) with a equal to the column of A that corresponds to the incoming variable (that characterizes the edge being traversed) and one or more sets of the form (1.5) have to be solved. It is also necessary to revise the product form to correspond to the replacement of the column leaving the basis by the incoming column a. A variant of the Bartels-Golub factorization (see Bartels,1971) which exploits sparsity will be described and compared with the algorithms of Forrest and Tomlin (1972) and Gill and Murray (1973). We will also describe a simple idea of Goldfarb (1975) that economizes the work in step (1.5)

At each step of the simplex method there is usually a considerable amount of choice over which simplex edge to follow (which column to bring into the basis). The standard choice (see Dantzig, 1963, page 159, for example) is to take the edge with best "reduced cost", that is best rate of improvement of the objective function with respect to the change in the out-of-basis variable that characterizes the edge. It has long been known (see Kuhn and Quandt, 1963, for example) that significantly less iterations are usually necessary if the steepest edge (along which the rate of improvement in E^n of the objective function is best) is always taken, but it has been assumed that it was not practicable to implement such an algorithm. Goldfarb has recently constructed recurrences that permit the implementation of this algorithm. We give these recurrences here and report the results of some experimental comparisons of the resulting algorithm with that of Harris (1973) and the standard algorithm.

2. Sparse Bartels-Golub algorithm

In this section we will describe our sparse variant of the Bartels-Golub algorithm. It begins with the application to the original basis of Gaussian elimination with row and column interchanges. It is convenient to express this elimination in the form of the equation

$$M_r M_{r-1} \cdots M_1 B = PUQ \tag{2.1}$$

where each M_i is a matrix which differs from I in just one off-diagonal element (and therefore represents an elementary row operation), P and Q are permutation matrices and U is upper-triangular. Equations (1.4) and (1.5) are each easy to solve since the factorisation (2.1) allows B^{-1} to be expressed in the form

$$B^{-1} = Q^T U^{-1} P^T M_r M_{r-1} \cdots M_1 \; . \tag{2.2}$$

After an iteration the new basis, $\overline{B}$ say, differs from B in just one column and so satisfies the equation

$$M_r M_{r-1} \cdots M_1 \, \overline{B} = PSQ \tag{2.3}$$

where S differs from U in just the one column in which B and $\overline{B}$ differ. Therefore S has the form illustrated in Figure 1. Bartels and Golub suggested using the column permutation that places the "spike" at the end and moves all intervening columns forward by one place to produce the upper Hessenberg form shown in Figure 2. Further row operations, perhaps including interchanges between adjacent rows, suffice to restore the upper triangular form. If these row operations are written as $M_{r+1},\ldots,M_{\bar{r}}$ then we have the new factorization

$$M_{\bar{r}}M_{\bar{r}-1}\ldots M_1\bar{B} = \bar{P}\,\bar{U}\,\bar{Q} \qquad (2.4)$$

which is exactly of the form (2.1).

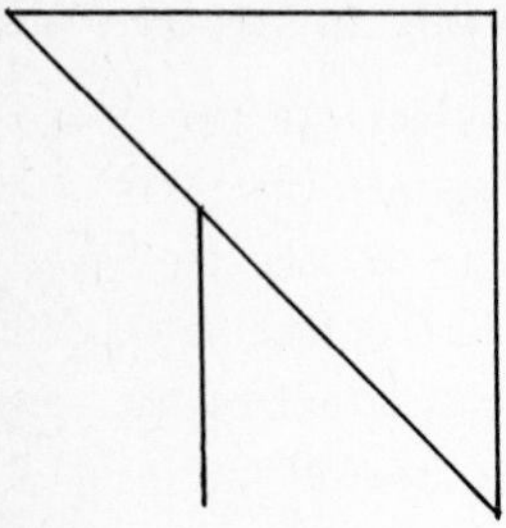

Figure 1

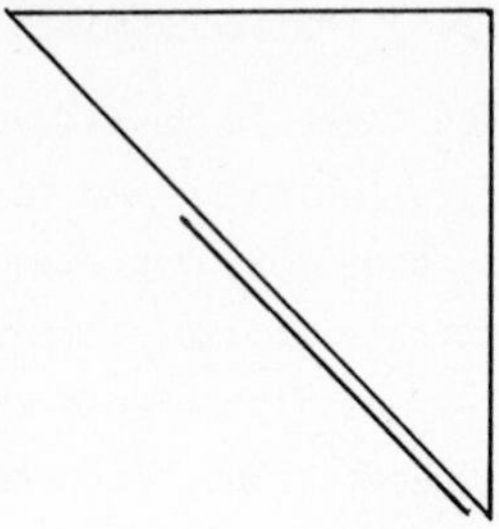

Figure 2

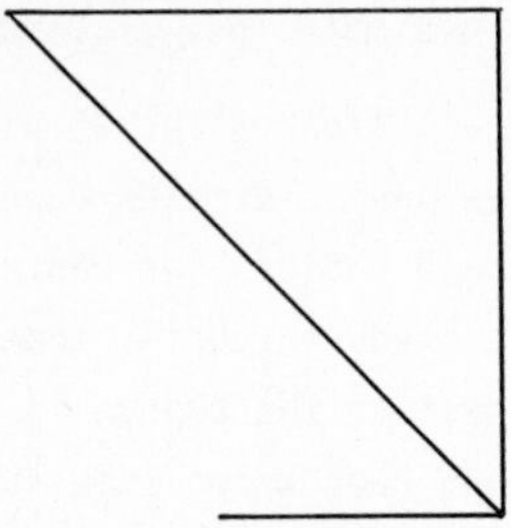

Figure 3

In the factorization (2.1) we want to maintain as much sparsity as possible and prefer non-zeros to be in U rather than in the product $M_r M_{r-1}\ldots M_1$ since a column of U is deleted at each simplex step whereas none of the matrices M_i is deleted. We have chosen Markowitz' (1957) pivotal strategy, so that each successive pivot is chosen to minimize the product of number of other non-zeros in its row with the number of other non-zeros in its column without regard to the effect that this choice may have on later pivotal steps. Where several potential pivots have identical "Markowitz cost" we bias the choice in favour of placing non-zeros in U rather than in matrices M_i. Markowitz' strategy was chosen because of our very happy experience with it elsewhere (see Duff and Reid, 1974).

To ensure numerical stability we take a non-zero to be a candidate for pivot only if it is greater than a multiple u of the largest element in its row of the current reduced matrix, where $u \leq 1$ is a preset parameter. Tomlin (1972) recommends the value 0.01 for u when using a pivotal strategy involving the assignment of the column order a priori from the sparsity pattern. The fill-in is less sensitive to the value of u when Markowitz' strategy is in use and we have found the more conservative value of 0.1 perfectly satisfactory. At each step of the reduction of the upper Hessenberg matrix we again use the parameter u, this time taking for pivot the larger of the two candidates unless they are within the factor u in modulus, in which case we take the one with least non-zeros in its row in order to minimize the fill-in produced by the step.

The key to stability in the eliminations is that the matrix elements must not grow drastically in size, for if they do this it must be because a large number is added to a small one and information present in the small one is lost. We therefore monitor the size of the largest element and refactorize the basis if this largest element becomes unreasonably large. Our experience is that such an event is very rare.

Saunders (private communication,1974) suggested that advantage could be taken of the fact that in the sparse case it is unlikely for the spike to extend to the last row of the matrix. If it extends to row ℓ (see Figure 4) then the column permutation that places the spike into column ℓ and moves the intervening columns forward by one place will give an upper Hessenberg matrix (see Figure 5) with less sub-diagonal elements.

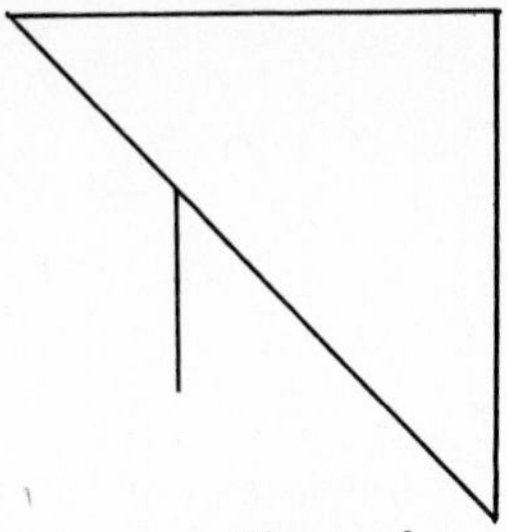

Figure 4

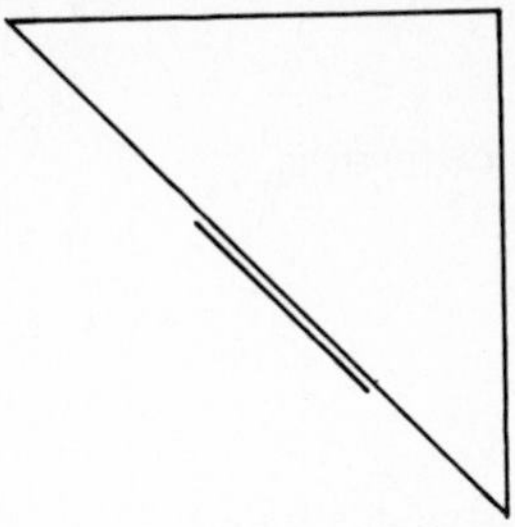

Figure 5

It is possible (and practicable) to improve Saunders' idea by applying further row and column permutations before any eliminations are performed. Details of how this is done **are given** by the author elsewhere (Reid,1975). Briefly, we work within "the bump" which consists of rows and columns s to ℓ if the spike is in column s and extends to row ℓ. If the bump contains a column singleton, other than the spike, it may be brought to the front of the bump by a symmetric permutation that preserves the form. The new bump, with order reduced by one, is treated similarly. All these permutations can in fact be generated in one sweep of the original bump. Next row singletons are treated similarly. Finally special action is taken if the spike is a singleton. The process as a whole has the effect of reducing the matrix to triangular form by permutations alone (and so without any fill-in) whenever this is possible.

Because the number of row operations M_i increases steadily and the number of non-zeros in U also on the whole increases it eventually becomes worthwhile to make an entirely fresh factorization of the current basis matrix. The frequency of re-factorization should depend on the rate of growth of numbers of non-zeros held and the relative speed of the original factorisation and the operations performed during a simplex step. In our early code (Reid,1973) we refactorize as soon as the number of non-zeros held reaches double the number it was at the previous refactorisation, but recently (Goldfarb and Reid,1975b) we have been following the advice of Beale (1971), to refactorize when the average time per iteration since last re-factorization begins to rise.

3. Backward transformation operations in the Bartels-Golub algorithm

The various product forms of the simplex method all require a backward transformation (solution of a set of equations (1.5)) in order to compute or revise the reduced costs. It can usually be arranged that this is relatively cheap.

Goldfarb's (1975) solution is to use the vector

$$q = \overline{B}^{-T} e_p, \tag{3.1}$$

where e_p is the column of the identity matrix corresponding to the spike, to update the reduced costs. In the sparse Bartels-Golub algorithm without the interchanges mentioned at the end of Section 2, the column permutation used implies that the relation

$$\overline{Q} e_p = e_m \tag{3.2}$$

holds, so that forming

$$\begin{aligned} q &= \overline{B}^{-T} e_p \\ &= M_1^T \dots M_r^T \overline{P}\, \overline{U}^{-T} \overline{Q} e_p \\ &= M_1^T \dots M_r^T \overline{P}\, \overline{U}^{-T} e_m \end{aligned} \tag{3.3}$$

does not involve a scan of U. If the additional interchanges are included, we find instead the relation

$$\overline{Q} e_p = e_\ell \tag{3.4}$$

where e_ℓ is a column of the identity matrix and is likely to be a late one. It follows that we need to form the vector

$$\overline{U}^{-T} e_\ell . \tag{3.5}$$

This involves a scan of just the part of U consisting of rows and columns ℓ to m, so is likely to be cheap.

4. Comparison of the two sparse Bartels-Golub algorithms with that of Gill and Murray

An algorithm which is similar to that of Bartels and Golub, but operates with orthogonal matrices instead of elementary Gaussian elimination matrices, was proposed by Gill and Murray (1973) and was adapted to the sparse case by Saunders (1972). It involves using the factorization

$$B = LQ \tag{4.1}$$

where L is lower-triangular and Q is orthogonal, but storing L only. Equations of the form (1.1) are solved with the aid of the identity

$$B^{-1} = B^T L^{-T} L^{-1}, \tag{4.2}$$

which may readily be verified from equation (4.1).

Data access is one of the most important contributors to computing costs, so we will use this to compare algorithms. Using (4.2) to solve (1.4) involves two accesses to L and one to B. Gill (1974) states that the rest of the algorithm requires two more accesses to L, making four in all. Our algorithm requires a forward and a backward scan of the file of matrices M_i and about two scans of U, depending on how the reduction of S goes and the sparsity of the vectors in (1.4) and (1.5). Thus we must expect the two algorithms to be roughly comparable if the number of non-zeros in the Murray-Gill-Saunders L is about half the number in the whole of the Bartels-Golub factorization.

Since orthogonal reductions cause more fill-in than Gaussian eliminations we must expect the matrix L to have more non-zeros than there are matrices M_i, but the fact that Q does not have to be stored is very helpful. Only numerical experiment can guide us. We therefore show in Table 1 a summary of some results obtained by Saunders (1972) using the Murray-Gill algorithm and the author using the sparse Bartels-Golub algorithm. Saunders used Algol-W on the IBM 360/91 and the author used the optimizing compiler (H with opt=2) on the IBM 370/165 for the 1973 runs and the IBM 370/168 for the 1975 runs. The 168 is about 25% faster than the 165 and the 91 is extrinsically faster still but it is the author's belief that its speed advantage is lost by the use of Algol.

The differences between the 1973 and 1975 results of the author are:

1) The objective function c is included in the basis in the 1973 case but not in 1975.
2) The 1973 results use a linked list for holding U whereas the 1975 ones use files in which have entries for each row and column (Gustavson,1972).
3) The additional row and column interchanges mentioned at the end of Section 2 are included in the 1975 results.
4) Some improvements to the code for the simplex method itself were included in 1975.
5) Reinversion was changed from when the total number of non-zeros in the factorized form doubled (1973) to when the average time since last reinversion began to increase.

Because in the Bartels-Golub algorithm the number of factors M_i increases steadily and the number of non-zeros in U also increases, Table 1 shows these numbers (always paired with the number of matrices M_i above the number of non-zeros in U) for the initial basis and before and after the last refactorization (except in the Shell case where the final numbers are shown since no reinversions were performed).

Saunders uses a different initial basis and handles simple bounds differently so the results are not strictly comparable. However it is absolutely clear that his algorithm does not perform as well on the last two cases as the sparse Bartels-Golub algorithm. On the first two the conclusions are not so clear-cut. In view of our remark earlier in this section that the algorithms are likely to be comparable if Saunders' L has half as many non-zeros as the Bartels-Golub factorization, let us compute ratios.

$$\rho = \frac{\text{average number of non-zeros in the Bartels-Golub form}}{\text{twice number of non-zeros in Saunders form}} .$$

For the 1973 Bartels-Golub code these ratios come to 1.2, 1.8, 0.3, 0.2 and for the 1975 results they come to 0.9, 1.0, 0.3, 0.2. We conclude that Saunders' results

Table 1

Problem	Shell			GUB			Stair			BP		
No. of rows, m		662			929			362			821	
No. of columns, n		1653			3333			544			1876	
No. of non-zeros in A		3310			10022			3843			10705	
No. of non-zeros in c		1344			3275			1			727	
	Saunders	Reid '73	Reid '75	Saunders	Reid '73	Reid '75	Saunders	Reid '73	Reid '75	Saunders	Reid '73	Reid '75
Initial non-zeros in factorized form	407	0 1687	0 1285	1428	0 3604	0 2714	2248	0 2454	0 2472	2263	0 3276	0 3014
Final non-zeros in factorized form	1046	1482 1902 0 1716	0 1286	1510	3415 3883 0 3609	53 3591 899 1832	12228	5401 3363 2026 3409	4957 4297 1998 3406	21000	4532 6075 2601 3957	2684 5395 1746 3206
Refactorizations	0	1	0	1	5	11	8	9	10	14	18	10
Refactorization time	0.18	1.1	0.3	0.23	2.8	.47	6.0	3.0	1.0	5.0	2.8	0.6
Iterations	343	367	259	958	1435	1956	488	591	631	284*	1836*	833*
Time per iteration (secs)	.07	.09	.05	.11	.30	.12	.75	0.18	0.12	1.0	0.35	0.11

*not converged

are better on the first two cases than the 1973 Bartels-Golub results, but the 1975 improvements make the algorithms extremely similar in performance. These remarks are confirmed by the computer times obtained.

It is interesting that the section 3 permutations improve the performance of the algorithm on the very sparse first two problems but help the other two hardly at all. Indeed the permutations are so successful on the Shell problem that no elimination matrices M_i are ever needed.

It is also apparent that there has been some gain in sparsity from having omitted c from the basis in the 1975 results, except in the case of Stair where c has only one non-zero. This can be judged by looking at the number of non-zeros in the initial factorized form and in the factorized form obtained after the last refactorization.

5. Comparison of the two sparse Bartels-Golub algorithms with that of Forrest and Tomlin

Another closely-related algorithm has been proposed by Forrest and Tomlin (1972). It differs from that of Bartels and Golub in the way the spiked matrix of Figure 1 is handled. The same permutation is used but is applied to the rows as well as the columns to produce the matrix shown in Figure 3. Next multiples of the earlier rows are added to the last one to eliminate the row spike.

Mathematically this algorithm is equivalent to reducing the upper Hessenberg matrix of Figure 2 to triangular form by eliminations between adjacent rows that always involve an interchange, but it is organised quite differently. A number of columns of U are quite unchanged, others are changed only by the removal of a single element, while only the last column is totally changed. Thus U may be held very conveniently by columns.

Because this algorithm corresponds to Bartels-Golub always with an interchange, we must expect some disadvantage in fill-in and stability. To assess these affects the author modified his 1973 sparse Bartels-Golub program, forcing it always to perform interchanges. The results of this test are summarized in Table 2. The worsened fill-in characteristics of the Forrest-Tomlin algorithm mean that more frequent refactorizations are needed, these being performed when the number of non-zeros exceeded double that after the previous refactorization. The worsened stability can be measured directly by looking at the largest matrix element or indirectly by looking at the in-basis reduced costs, which should all be zero. It may be seen that the Forrest-Tomlin algorithm is inferior on both fill-in and stability, but not sufficiently to preclude its use where its organisational advantages are wanted. In any case the possible instability may easily be avoided by monitoring the growth in the size of matrix elements and performing a refactorization when necessary.

Table 2

	Shell		GUB		Stair		BP	
	F/T	B/G	F/T	B/G	F/T	B/G	F/T	B/G
Average frequency on refactorization	172	318	134	399	72	71	117	130
Maximum matrix element	1	1	10^3	20	3×10^7	3×10^4	5×10^4	4×10^2
Maximum in-basis reduced cost	$<10^{-15}$	$<10^{-15}$	10^{-9}	3×10^{-10}	6×10^{-6}	6×10^{-8}	4×10^{-8}	10^{-7}

Comparisons were also made using our 1975 code. For these experiments we included a problem, kindly supplied by S. Powell, which has 548 rows, 1076 columns and 7131 non-zeros. The results are summarized in Table 3. The first four problems were timed to completion but BP was terminated about a third of the way through phase 2 at the same value of the objective function. The new algorithm was very successful in the first three problems, particularly on Shell where no eliminations were ever necessary. On Powell and GUB the number of iterations with no eliminations was 176 (out of 226) and 967 (out of 1119). On the last two, rather denser, problems the algorithms showed very similar performance.

Table 3

	Average interval between refactorizations		Time	
	SBG	F/T	SBG	F/T
Powell	111	45	17.7	19.3
Shell	∞	∞	15.5	17.2
GUB	120	79	174	202
Stair	46	51	46.9	45.8
BP	64	64	289	297

6. The steepest-edge simplex algorithm

We consider a single step of the simplex method, using bars to distinguish quantities after the step from those present at its start, and suppose that the variables are ordered so that the first m are basic at the beginning of the step.

If the iteration involves bringing column q (q>m) into the basis in place of column p (p≤m) then the new iterate is

$$\bar{x} = x + \theta\eta_q, \quad \eta_q = \begin{pmatrix} B^{-1}a_q \\ e_q \end{pmatrix}, \tag{6.1}$$

where θ is a scalar chosen so that $\bar{x}_p=0$. It is usual Dantzig,1963) to choose the column which minimizes the "reduced cost"

$$z_q = c^T \eta_q, \tag{6.2}$$

which is the change in the objective function per unit change in x_q. We consider instead the alternative of minimizing the normalized reduced cost

$$c^T \eta_q / ||\eta_q||_2 \tag{6.3}$$

which corresponds to choosing the edge of steepest descent in E^n of the objective function.

For large problems, explicit computation of all the norms $||\eta_i||_2$, $i>m$, at each step is prohibitively expensive. Fortunately Goldfarb has shown (see Goldfarb and Reid,1975a) that the well-known Sherman-Morrison formula (see Householder,1964, page 123, for example) may be used to yield the updating formulae

$$\bar{\eta}_q = \eta_q / \alpha_q \tag{6.4a}$$

$$\bar{\eta}_i = \eta_i - \eta_q \bar{\alpha}_i \;, \qquad i>m,\ i\neq q \tag{6.4b}$$

where α_q is the pth element of $B^{-1}a_q$ and the numbers $\bar{\alpha}_i$, $i>m$, are the components of row p of $\bar{B}^{-1}A$, i.e.

$$\bar{\alpha}_i = (\bar{B}^{-T} e_p)^T a_i \;. \tag{6.5}$$

This enables the recurrences

$$\bar{\gamma}_q = \gamma_q / \alpha_q^2$$

$$\bar{\gamma}_i = \gamma_i - 2\bar{\alpha}_i a_i^T B^{-T} B^{-1} a_q + \bar{\alpha}_i \gamma_q, \quad i\neq q \tag{6.6b}$$

for the numbers

$$\gamma_i = ||\eta_i||_2^2 = \eta_i^T \eta_i \tag{6.7}$$

to be constructed. The vector $\bar{B}^{-1}e_p$ in (6.5) may be calculated economically in view of our remarks in Section 5, and may be used to update the reduced costs z_i (see (6.2)). The vector $B^{-1}a_q$ is needed anyway (see (6.1)), but extra work is involved in calculating $B^{-T}(B^{-1}a_q)$ from it by a backward transformation operation. Extra storage is of course needed for the numbers γ_i, but little arithmetic is likely to be involved in implementing (6.6b) since most of the numbers $\bar{\alpha}_i$ are usually zero in practical problems.

The algorithm of Harris (1973) can be regarded as a variant in which the weights γ_i are calculated approximately. She takes weights T_i which approximate the semi-norms $||\underline{\eta}_i||_h$ obtained by using the Euclidean norm of the subvector consisting of just those components in the current "reference framework". She takes the initial out-of-basis variables to constitute the initial reference framework and periodically revises the framework to become the set of current out-of-basis variables. Initially, at each such revision, all her weights are exactly one. At other iterations she updates them by the formulae

$$\bar{T}_q = \max(1, ||\eta_q||_h/\alpha_q) \qquad (6.8a)$$

$$\bar{T}_i = \max(T_i, ||\eta_q||_h\, \bar{\alpha}_i), \quad i \neq q . \qquad (6.8b)$$

The vector η_q is calculated explicitly, as in equation (6.1) of our algorithm and so its semi-norm may be evaluated. Equation (6.4a) suggests that $\bar{T}_q$ should be $||\eta_q||_h/\alpha_q$ but she uses (6.8a) to avoid very small weights which may result if variable q is not in the current reference framework. Formula (6.8b) is obtained directly from (6.4b) by using the larger of the norms of two vectors as an approximation to the norm of their sum. It is clear from (6.8b) that weights T_i associated with out-of-basis variables that remain out-of-basis increase steadily and it is therefore necessary to reinitialize the reference framework (and reset the weights to unity) from time to time. The algorithm is a little more economical than the steepest edge algorithm because the vector $B^{-T}(B^{-1}a_q)$ is not needed and inner products of this vector with non-basic columns of A in (8.6b) are not required. The latter saving is typically not very great because sparsity ensures that most $\bar{\alpha}_i$ are zero. There is no gain in storage.

7. Practical tests of column pivoting algorithms

In this section we report the results of using a slightly modified version of our code (Goldfarb and Reid, 1975b) on six test problems. The modifications were incorporated so that we could run the original Dantzig algorithm and that of Harris as well as the steepest edge algorithm. We have tried to write the code efficiently so that realistic timings could be made. The presence of sparsity means that there is no way to predict from the recurrences themselves just how expensive each will be and it seems best to rely on actual machine times.

The test problems were kindly provided by P. Gill (Blend), S. Powell (Powell) and M.A. Saunders (Stair, Shell, GUB and BP). Each problem was begun in the same way with a primitive "crash" code that generates an initial basis which is a permutation of a triangular matrix. In order that they should all have the same starting basis for phase 2 we used the feasible point generated by the steepest edge algorithm to restart all three algorithms.

Our results are summarized in Table 4. In the case of Blend our "crash" routine generated a basis that gave a feasible solution so comparisons could be made only on phase 2. In the case of Powell, the problem had no feasible solution so we could only compare algorithms in phase 1. The others had two phases and we ran each algorithm on each phase, apart from the original Dantzig algorithm on BP which was curtailed because of the high computing cost.

We have given separately the time taken to calculate γ_i, $i>m$, initially for the steepest-edge algorithm. This is an overhead only for phase 1 and it can be seen that it is usually quite slight compared with the rest of the time taken in phases 1 and 2, the worst overhead being 18% on Shell and the average being 5%.

Table 4 Summary of results

Problem			Blend	Powell	Stair		Shell		GUB		BP	
Number of rows, m			74	548	362		662		929		821	
Number of cols, n			114	1076	544		1653		3333		1876	
Number of non-zeros			560	7131	4013		5033		14158		11719	
Phase			2	1	1	2	1	2	1	2	1	2
Iterations	Steepest edge		48	226	210	157	48	183	475	644	1124	1182
	Harris		65	198	313	163	77	181	672	796	1414	1819
	Dantzig		70	285	244	387	78	181	854	1102	Not run	>3976
Time	Steepest edge	Set up	0.02	1.1	0.4	-	2.4	-	7.5	-	3.5	-
		run	0.90	16.6	26	28	2.9	10.2	57	109	190	263
	Harris		1.04	11.7	33	24	3.8	8.6	78	111	195	305
	Dantzig		1.06	16.9	26	51	3.8	8.5	96	146	Not run	>596
Time per iteration	Steepest edge		.019	.073	.12	.18	.060	.056	.12	.17	.17	.22
	Harris		.016	.059	.11	.15	.049	.048	.12	.14	.14	.17
	Dantzig		.015	.059	.11	.13	.049	.047	.11	.13	Not run	.15

In terms of numbers of iterations, the steepest-edge algorithm was best except for Powell (14% more iterations than Harris) and phase 2 of Shell (1% more iterations than Harris) and the average percentage gain over the Harris algorithm was 19%. On the other hand the time per iteration was on average 22% greater, so that overall the two algorithms are very comparable. In fact the total time taken by the steepest-edge algorithm was less than that taken by the Harris algorithm on Blend, GUB and BP and greater on Powell, Stair and Shell.

Both the steepest-edge and the Harris algorithms show a worthwhile overall gain over the original Dantzig algorithm, being significantly better on Stair, GUB and BP and comparable on the others. This is further illustrated by the experience of Harris (1973, Figures 5,6,7).

8. Conclusions

We have demonstrated that it is practicable to implement a variant of the Bartels-Golub algorithm that sometimes avoids the need for any eliminations (and fill-ins) at an iteration and always does so if this is possible by permutations alone. The algorithm has better stability and fill-in properties than the closely related algorithm of Forrest and Tomlin (1972), although numerical experiments indicate that these advantages are usually quite slight in practice. When compared with the stable algorithm of Gill and Murray (1973), implemented by Saunders (1972) it was shown to require a similar amount of computation for two (very sparse) cases but much less on two (rather fuller) cases.

We have also shown that it is practicable to implement the steepest-edge column selection algorithm. Overall results with it appear to be very comparable with those obtained with Harris' algorithm, both sometimes showing very worthwhile gains over the standard algorithm.

9. Acknowledgements

The author wishes to thank P.E. Gill, D. Goldfarb, M.A. Saunders and J.A. Tomlin for their helpful comments in discussion and P.E. Gill, S. Powell and M.A. Saunders for providing test problems.

References

Bartels, R.H. (1971). A stabilization of the simplex method. Num. Math., 16, 414-434.

Beale, E.M.L. (1971). Sparseness in linear programming. In "Large sparse sets of linear equations". Ed. J.K. Reid, Academic Press.

Dantzig G.B. (1963). Linear programming and extensions. Princeton University Press.

Duff, I.S. and Reid, J.K. (1974). A comparison of sparsity orderings for obtaining a pivotal sequence in Gaussian elimination. J. Inst. Maths. Applics., 14, 281-291.

Forrest, J.J.H. and Tomlin, J.A. (1972). Updating triangular factors of the basis to maintain sparsity in the product form simplex method. Mathematical programming, 2, 263-278.

Gill, P.E. and Murray, W. (1973). A numerically stable form of the simplex algorithm. Linear Alg. Appl., 7, 99-138.

Goldfarb, D. (1975). On the Bartels-Golub decomposition for linear programming bases. To appear.

Goldfarb, D. and Reid, J.K. (1975a). A practical steepest edge simplex algorithm. To appear.

Goldfarb, D. and Reid, J.K. (1975b). Fortran subroutines for sparse in-core linear programming. A.E.R.E. Report to appear.

Gill, P.E. (1974). Recent developments in numerically stable methods for linear programming. Bull. Inst. Maths. Applics. 10, 180-186.

Gustavson, F.G. (1972). Some basic techniques for solving sparse systems of linear equations. In "Sparse matrices and their applications". Ed. D.J. Rose and R.A. Willoughby, Plenum Press.

Harris, P.M.J. (1973). Pivot selection methods of the Devex LP code. Mathematical programming 5, 1-28.

Householder, A.S. (1964). The theory of matrices in numerical analysis. Blaisdell.

Kuhn, H.W. and Quant, R.E. (1963). An experimental study of the simplex method. Proc. of Symposia in Applied Maths, Vol.XV, Ed. Metropolis et al. A.M.S.

Markowitz, H.M. (1957). The elimination form of the inverse and its applications to linear programming. Management Sci., 3, 255-269.

Reid, J.K. (1973). Sparse linear programming using the Bartels-Golub decomposition. Verbal presentation at VIII International Symposium on Mathematical Programming, Stanford University.

Reid, J.K. (1975). A sparsity-exploiting variant of the Bartels-Golub decomposition for linear programming bases. To appear.

Saunders, M.A. (1972). Large-scale linear programming using the Cholesky factorization. Report STAN-CS-72-252, Stanford University.

Tomlin, J.A. (1972). Pivoting for size and sparsity in linear programming inversion routines. J. Inst. Maths. Applics., 10, 289-295.

TOWARDS A THEORY FOR DISCRETIZATIONS OF STIFF DIFFERENTIAL SYSTEMS

Hans J. Stetter

I. SINGULARLY PERTURBED SYSTEMS

In the study of discretizations of stiff systems, "models" have played a big role from the beginning. The (scalar) model equation

(1.1) $$y' = \lambda y, \qquad \lambda \in \mathbb{C},\ \mathrm{Re}\lambda < 0$$

has led to the concept of stability regions and A-stability, L-stability, etc; but it is too simple. The (vector) model equation

(1.2) $y' = f(t,y)$ with an exponentially stable equilibrium,

was exploited to some extent in [1]; however, it does not permit a distinction between solution components of different growth rates.

A more refined model should possess the following properties:

a) It should permit the simultaneous occurence of slowly varying and of rapidly decaying solution components.

b) It should permit the consideration of a limit process corresponding to a transition to arbitrarily high stiffness. (Such a limit process would facilitate the analysis and lead to concepts which should prove useful also in non-limit situations.)

A model with these features has been proposed by various authors during the past years (see, e.g., [2], [3], [4]), viz. singularly perturbed systems of ordinary differential equations:

(S) $$\begin{aligned} x' &= \frac{1}{\varepsilon} f(t,x,y,\varepsilon), \qquad \mu[\frac{\partial}{\partial x} f] < 0 \text{ in a suitable region,} \\ y' &= g(t,x,y,\varepsilon); \end{aligned}$$

$\varepsilon > 0$ is a small parameter, $x(t) \in \mathbb{R}^{s_1}$, $y(t) \in \mathbb{R}^{s_2}$, $s_1 \geq 1$, $s_2 \geq 0$; μ is the logarithmic norm of the matrix $\frac{\partial}{\partial x} f$. An initial value problem on some interval $[0,T]$ is considered, with initial values x_0, y_0.

Some of the mathematical theory for (S) may be found in [5], chapters 39 and 40: Any solution of (S) permits an asymptotic expansion in ε of the type $(R \geq 1)$

$$\begin{pmatrix} x(t,\varepsilon) \\ y(t,\varepsilon) \end{pmatrix} =: z(t,\varepsilon) = \sum_{r=0}^{R-1} \tilde{z}_r(t)\varepsilon^r + \sum_{r=0}^{R-1} \hat{z}_r(\tfrac{t}{\varepsilon})\varepsilon^r + O(\varepsilon^R)$$

$$(1.3) \qquad\qquad =: \quad \tilde{z}(t,\varepsilon) \quad + \quad \hat{z}(t,\varepsilon) \quad + O(\varepsilon^R);$$

here both the $\tilde{z}_r$ and $\hat{z}_r$ are well-behaved functions, and $\hat{z}_r(\tau) \to 0$ as $\tau \to \infty$ for $r = 0(1)R-1$.

Thus, except for a remainder term of order $O(\varepsilon^R)$, $R > 0$, a solution of (S) may be uniquely decomposed into

- a smooth or regular component $\tilde{z}$ which may be characterized by the fact that its derivatives remain bounded as $\varepsilon \to 0$;
- a fast or singular component $\hat{z}$ which is characterized by a decay which becomes the more rapid the smaller ε gets; normally $\|z(t)\| \leq c \exp(-\frac{m}{\varepsilon}t)$, with some $m > 0$.

This decomposition (1.3) of the solution z of an initial value problem for (S) implies a decomposition of the given initial value $z_o = \begin{pmatrix} x_o \\ y_o \end{pmatrix}$:

$$(1.4) \qquad z_o = \tilde{z}(0,\varepsilon) + \hat{z}(0,\varepsilon) + O(\varepsilon^R) =: \tilde{z}_o(\varepsilon) + \hat{z}_o(\varepsilon) + O(\varepsilon^R)$$

The prescription of $\tilde{z}_o(\varepsilon)$ in place of z_o would lead to the same regular component $\tilde{z}(t,\varepsilon)$, with $\hat{z}(t,\varepsilon) \equiv 0$.

Trivial example: $x' = -\frac{1}{\varepsilon}x, \qquad y' = x$

$$z(t,\varepsilon) = \begin{pmatrix} 0 \\ y_o + \varepsilon x_o \end{pmatrix} + \begin{pmatrix} x_o \exp(-t/\varepsilon) \\ -\varepsilon x_o \exp(-t/\varepsilon) \end{pmatrix},$$

$$\tilde{z}_o(\varepsilon) = \begin{pmatrix} 0 \\ y_o + \varepsilon x_o \end{pmatrix}, \qquad \hat{z}_o(\varepsilon) = \begin{pmatrix} x_o \\ -\varepsilon x_o \end{pmatrix}$$

Here, the y-component of the regular solution is a constant whose value depends also on the initial value x_o of the x-component and on ε.

II. DISCRETIZATIONS OF (S)

We will consider discretizations of (S) on grids characterized by a step parameter h and we will normally let ε tend to zero for fixed $h > 0$. In some cases it may also be interesting to consider a simultaneous decrease of ε and h, with h behaving like $O(\varepsilon)$, $O(\sqrt{\varepsilon})$, etc. However, the quantity

h/ε will always remain bounded away from zero. (The limit process $h/\varepsilon \to 0$ would take us back into the realms of the well-known Dahlquist-Henrici theory of discretizations on "sufficiently fine" grids.)

Thus, our two parameters ε and h will vary in the trapezoidal region ($c > 0$ is an arbitrary factor)

(2.1) $\quad E := \{(h,\varepsilon) \in \mathbb{R}^2 : \varepsilon \in (0,\varepsilon_0], \quad c\varepsilon \leq h \leq h_0\}$

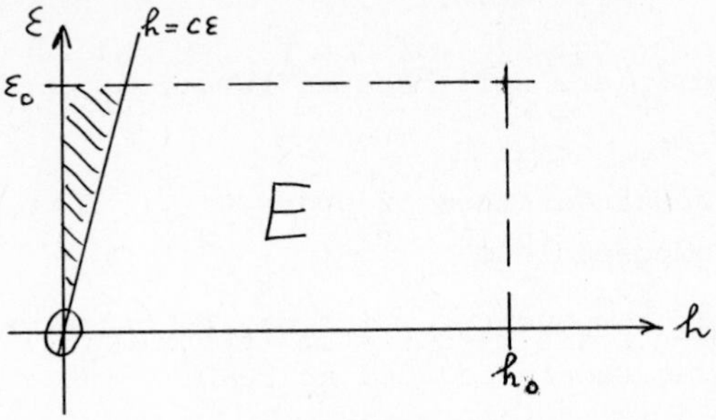

Our concepts should be such that they hold uniformly in E, i.e. uniformly in ε, $\varepsilon \in (0,\varepsilon_0]$, for all sufficiently small $h \geq c\varepsilon$. Quantitative concepts will be based on limits $\varepsilon \to 0$, with $h > 0$ fixed or $h = O(\varepsilon^\alpha)$, $0 < \alpha \leq 1$.

A solution of a discretization of a system (S) with some given initial value z_0 will be denoted by $\zeta(h,\varepsilon)$, its value at some gridpoint t_n by $\zeta_n(h,\varepsilon)$. The following are natural requirements for such solutions in the context of our model:

1) Boundedness: $\zeta(h,\varepsilon)$ should be bounded uniformly in E. Here we have assumed that the basic interval $[0,T]$ is such that the solution of (S) exists and is bounded in $[0,T]$ for all $\varepsilon \in (0,\varepsilon_0]$.

2) R-S-Decomposibility: $\zeta(h,\varepsilon)$ should be uniquely decomposible into a regular (smooth) and a singular (fast) component

(2.2) $\quad \zeta_n(h,\varepsilon) = \tilde{\zeta}_n(h,\varepsilon) + \hat{\zeta}_n(h,\varepsilon) + O(\varepsilon^R), \quad R \geq 1.$

Again, the regular component may be characterized by the fact that its difference quotients (up to some order r) are bounded uniformly in E.

The singular component must decay with increasing t; but we will normally be able to achieve only a decay like

(2.3) $\quad \|\hat{\zeta}_n(h,\varepsilon)\| \leq c\, q^n, \qquad q < 1, \quad$ uniformly in E.

In favorable cases, we will have $q = O(\varepsilon)$. In any case, this will mean that the difference quotients of $\hat{\zeta}$ are not bounded uniformly in E. Note that this last property is also shared by components which oscillate without a decay in amplitude. Hence we might also admit such singular components (which can be eliminated by smoothing).

The R-S-decomposition (2.2) of ζ implies an R-S-decomposition of the initial value ζ_o - or of the initial values $\boldsymbol{\zeta}_o := (\zeta_o, \zeta_1, \ldots, \zeta_{k-1})$ in case of a k-step method:

(2.4) $$\boldsymbol{\zeta}_o(h,\varepsilon) = \tilde{\boldsymbol{\zeta}}_o(h,\varepsilon) + \hat{\boldsymbol{\zeta}}_o(h,\varepsilon) + O(\varepsilon^R),$$

where $\boldsymbol{\zeta}_o(h,\varepsilon)$ is determined by z_o and the discretization method.

3) R-convergence: The regular component $\tilde{\zeta}$ of ζ should approximate the regular component $\tilde{z}$ of the solution z of (S) uniformly in E and sufficiently well for small h:

(2.5) $$\sup_{\varepsilon\in[0,\varepsilon_o],h=\text{const}} \|\tilde{\zeta}_n(h,\varepsilon) - \tilde{z}(t_n,\varepsilon)\| = E(t_n,h) = O(h^p)$$

while no requirements are made for an approximation of the singular component $\hat{z}$ by $\hat{\zeta}$.

This concept is based on the assumption that it is the smooth component $\tilde{z}$ only in which we are interested, ε being so small that the rapid transient $\hat{z}$ will not contribute anything appreciable after a very short time interval. It also assumes that no effort is made to reproduce the transient phase of the solution in the discretization but that one is satisfied to have a good reproduction of the stationary phase (although only from some $\bar{t}$ onwards). In this sense, R-convergence is related with Gear's concept of stiff stability (see [6]).

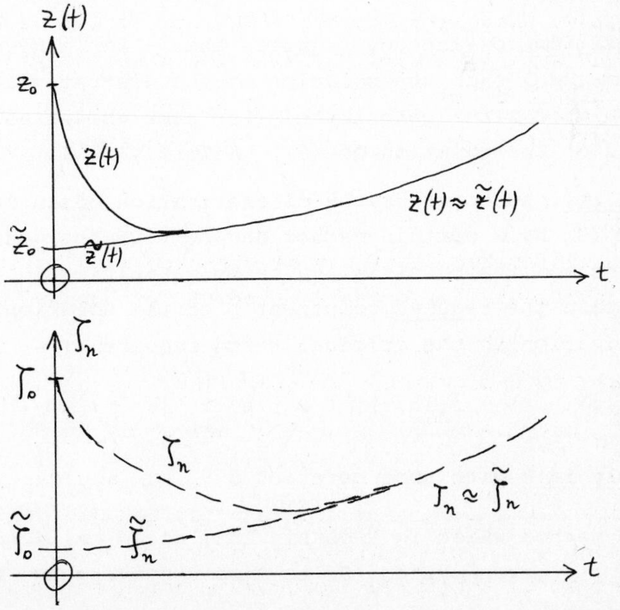

In some cases it will be more reasonable to consider - in place of (2.5) - a relative measure of error: If $\tilde{z}(t,\varepsilon) = O(\varepsilon)$ in some components $\tilde{z}_s$ then one should have $\tilde{\zeta}_n(h,\varepsilon) = O(\varepsilon)$ in those components and also $|\tilde{\zeta}_{n,s}(h,\varepsilon) - \tilde{z}_s(t_n,\varepsilon)|/\varepsilon = O(h^p)$.

Auxiliary Concepts:

In the classical "$h \to 0$" theory of discretizations the concepts of consistency and D-stability have been introduced as means for establishing convergence (see, e.g., the discussion in [1], section 2.2). The following concepts which are adapted from concepts in [4] may prove useful in the same way for R-convergence.

1) <u>R-consistency</u>: We form the local discretization error for the regular component $\tilde{z}$ of the solution z of (S) as the residual obtained by inserting $\tilde{z}$ into the difference equation (in normal form, see [1], section 1.1); we require this quantity $\tilde{l}(t;h,\varepsilon)$ to be bounded uniformly in E and to be small for small h:

$$\sup_{\varepsilon \in (0,\varepsilon_0]} \|\tilde{l}(t;h,\varepsilon)\| = L(t,h) = O(h^q), \quad q > 0.$$

2) <u>Decomposition consistency</u>: We consider the values of $\tilde{z}$ and $\tilde{\zeta}$ at $t_0,\ldots,t_{k-1}$ (for a k-step method) which are generated by a given initial value z_0 for (S) and a certain method; we require

$$\sup_{\varepsilon \in (0,\varepsilon_0]} \|\tilde{\zeta}_0(h,\varepsilon) - \tilde{z}_0(\varepsilon)\| = Z(h) = O(h^{q_0}), \quad q_0 > 0.$$

Decomposition consistency requires that - for a given z_0 - the regular component $\tilde{\zeta}$ of the solution of the discretization should start on the h-grid with approximately the same values as the regular component $\tilde{z}$ of the solution of (S), uniformly in E.

3) <u>R-stability</u>: We consider the discretization of an initial value problem for (S) by a certain method and perturb the generated difference equations and their initial value by quantities δ_n and δ_0 resp.; we now compare the <u>regular</u> component $\tilde{\zeta}'$ of the solution of this perturbed discretization to the original $\tilde{\zeta}$ and require

$$\|\tilde{\zeta}'_n(h,\varepsilon) - \tilde{\zeta}_n(h,\varepsilon)\| \leq S\,[\,h \sum_{\nu=k}^{n} \|\delta_\nu\| + \|\delta_0\|\,]$$

<u>uniformly in E</u> with some constant S.

<u>Theorem</u>: A method which is R-consistent, decomposition consistent and R-stable for a given system (S) is also R-convergent, with $p = \min(q,q_0)$

Proof (idea): $\tilde{z}$ satisfies the system of difference equations generated by the discretization except for a "perturbation" $\tilde{l}(t_n;h,\varepsilon)$. The initial values of the two systems of difference equations differ by $-[\tilde{\zeta}_o(h,\varepsilon)-\tilde{z}_o(\varepsilon)]$. The difference quotients (up to some order r) of $\tilde{z}$ are bounded uniformly in E since the derivatives (up to some order r) of $\tilde{z}$ are bounded uniformly in ε; hence $\tilde{z}$ must be the regular component of the solution of the perturbed system. Now, the assumptions imply the uniform smallness of the difference between $\tilde{z}$ and $\tilde{\zeta}$.

This theorem seems to indicate that the auxiliary concepts introduced above have been chosen reasonably; however, this could only be claimed if the establishment of R-consistency, decomposition consistency, and R-stability in a given situation is easier than the immediate establishment of R-convergence. Whether this is the case is not yet clear.

III. FURTHER ANALYSIS

So far, reasonably general sufficient conditions do not even exist for the boundedness (uniform in E, see section II) of the solutions of a discretization of (S). Examples have shown that A-stability is not sufficient, see [1], section 3.5. "Strong A-stability" (i.e. ∞ lies in the interior of the region of absolute stability) may be sufficient under mild restrictions on (S). In [7], Dahlquist proves boundedness (convergence) for a certain class of multistep methods and a certain class of problems under a condition called G-stability.

The well-known construction principle for the R-S-decomposition (1.3) of a solution of (S) (see, e.g., [3]) unfortunately cannot be extended to discretizations of (S): With (1.3), $\hat{x}_r'(\frac{t}{\varepsilon}) = O(\varepsilon^{-1})$ and thus relates to $\hat{x}_{r-1}$ in the asymptotic expansion. An analogous property for the singular part of the difference solution cannot be expected.

Consider a strongly A-stable method and a system (S) where the $\frac{1}{\varepsilon}$ affects only linear terms with constant coefficients; then (see [4]) one has a quantity $\rho(h,\varepsilon) < 1$ such that a decomposition (2.2) exists with

$$\|\hat{\zeta}_n(h,\varepsilon)\| \leq c\,[\rho(h,\varepsilon)]^n \leq \bar{c}\,[\rho(h,\varepsilon)]^{t_n/h} \quad (3.1)$$

In the following simple example, this decomposition can be constructed explicitly:

Example: $x' = -\frac{1}{\varepsilon}x + \frac{1}{\varepsilon}e^t$ (there is no "smooth" part, i.e. $s_2 = 0$)

$\tilde{x}(t) = \frac{e^t}{1+\varepsilon}$ for arbitrary x_o.

The implicit Euler-method yields, with a constant step h,

$$\xi_n = \xi_{n-1} - \frac{h}{\varepsilon}\xi_n + \frac{h}{\varepsilon} e^{t_n}$$

$$\xi_n = \frac{1}{1+\frac{h}{\varepsilon}}\xi_{n-1} + \frac{h}{\varepsilon(1+\frac{h}{\varepsilon})} e^{t_n} = \frac{\varepsilon}{\varepsilon+h}\xi_{n-1} + \frac{h}{\varepsilon+h} e^{t_n}$$

$$\xi_n = \left(\frac{\varepsilon}{\varepsilon+h}\right)^n \xi_o + \frac{h}{\varepsilon+h}\sum_{\nu=1}^{n}\left(\frac{\varepsilon}{\varepsilon+h}\right)^{n-\nu} e^{\nu h}$$

$$(3.2) \qquad = \left(\frac{\varepsilon}{\varepsilon+h}\right)^n \left[\xi_o - \frac{1}{1+\varepsilon\frac{1-e^{-h}}{h}}\right] + \frac{e^{t_n}}{1+\varepsilon\frac{1-e^{-h}}{h}} = \hat{\xi}_n(h,\varepsilon) + \tilde{\xi}_n(h,\varepsilon)$$

Here $\rho(h,\varepsilon) = \frac{\varepsilon}{\varepsilon+h} < 1$ in E.

It may also be reasonable to consider decompositions where the "singular part does not decay rapidly but oscillates rapidly. Such a behavior occurs with methods where the "growth factor" per step tends to -1 with $\varepsilon \to 0$ for the test equation $x' = -\frac{1}{\varepsilon}x$. Such oscillations also permit a unique identification of the "singular" component. Although it does not decay it can be smoothed away the better the smaller ε is.

Example: Differential equation as above.
The implicit midpoint method

$$\xi_n = \xi_{n-1} - \frac{h}{\varepsilon}\frac{\xi_n + \xi_{n-1}}{2} + \frac{h}{\varepsilon} e^{t_n - \frac{h}{2}}$$

yields after some manipulation

$$(3.3) \quad \xi_n = \left(-\frac{h-2\varepsilon}{h+2\varepsilon}\right)^n \left[\xi_o - \frac{1}{\cosh(h/2)+\varepsilon\frac{\sinh(h/2)}{h/2}}\right] + \frac{e^{t_n}}{\cos(h/2) + \varepsilon\frac{\sin(h/2)}{h/2}}$$

After symmetric smoothing, $\hat{\xi}_n$ reduces to

$$\hat{\xi}'_n = \frac{1}{4}\left(\hat{\xi}_{n-1} + 2\hat{\xi}_n + \hat{\xi}_{n+1}\right) = -\frac{4\varepsilon^2}{h^2 - 4\varepsilon^2}\left(-\frac{h-2\varepsilon}{h+2\varepsilon}\right)^n \left[\xi_o - \frac{1}{\cos(h/2) + \varepsilon\frac{\sinh(h/2)}{h/2}}\right]$$

which is $O(\varepsilon^2)$. Thus, if (2.2) is only carried to R = 2, the smoothed solution of the discretization possesses an R-S-decomposition proper, with a vanishing singular component.

From the explicit R-S-decompositions in the two examples above, we can immediately deduce R-convergence: The regular component of the solution (3.2) (implicit Euler method) satisfies

$$\tilde{\xi}_n - \tilde{x}(t_n) = \frac{e^{t_n}}{1+\varepsilon\frac{1-e^{-h}}{h}} - \frac{e^{t_n}}{1+\varepsilon} = e^{t_n}\frac{\varepsilon h}{2}\,(1+O(h)+O(\varepsilon));$$

with the implicit midpoint method (cf. (3.3)) we obtain

$$\tilde{\xi}_n - \tilde{x}(t_n) = -\,e^{t_n}\,h^2\left(\frac{1}{8} - \frac{5}{24}\varepsilon + O(h^2) + O(h^2\varepsilon)\right)$$

and

$$\tilde{\xi}'_n - \tilde{x}(t_n) = e^{t_n}\,h^2\left(\frac{1}{8} - \frac{1}{24}\varepsilon + O(h^2) + (h^2\varepsilon)\right).$$

These expressions show a difference in structure: While

$$\lim_{\varepsilon\to 0} |\tilde{\xi}_n - \tilde{x}(t_n)| = 0 \qquad \text{for fixed } h > 0$$

for the implicit Euler method, the same quantity is $O(h^2)$ for the midpoint method.

This suggests a characterization of the <u>order of R-convergence</u> by <u>two</u> natural numbers: We will say that a given method is R-convergent of order (p,t) for a given problem if

(3.5) $$\|\tilde{\zeta}_n - \tilde{z}(t_n)\| = O(h^p\varepsilon^t) \qquad \text{as } h\to 0 \text{ and } \varepsilon\to 0 \text{ \underline{independently}}.$$

Such a notion has been introduced similarly by Prothero and Robinson ([8]). It depends, however, strongly on the problem under consideration; the numbers p and t may also differ for the individual vector components of a vector-valued solution.

<u>Example</u>: If we replace the differential equation of the previous examples by

$$x' = -\frac{1}{\varepsilon}\,x + e^t$$

then $\tilde{x}$ and $\tilde{\xi}$ are multiplied by ε and t is increased by one!

IV. SIMPLER MODELS

Initially, it is advisable to consider simpler models than (S). Hopefully, the results and techniques found in the study of these models will lead to a successful treatment of the general system (S).

The following model has been used by Prothero and Robinson in their paper [8] (we have adapted the notation):

(4.1) $$z' = \frac{\lambda}{\varepsilon}\,(z - g(t)) + g'(t), \qquad \mathrm{Re}\,\lambda < 0.$$

It is obvious that

$$\tilde{z}(t) = g(t) \qquad \text{for all } z_o,\ \hat{z}(t) = (z_o - g(o))e^{\frac{\lambda}{\varepsilon}t}.$$

In [8], the authors introduce a concept of S-stability for one-step methods based upon this model; it is related to our concept of R-convergence.

The model (4.1) has two serious limitations:

a) (4.1) is linear with a constant coefficient.

b) The substitution $x(t) := z - g(t)$ transforms (4.1) into the trivial model equation

(4.2) $$x' = \frac{\lambda}{\varepsilon} x.$$

Application of the same substitution to the discretization of (4.1) can lead to the corresponding discretization of (4.2) only if the method samples the right hand side at the gridpoints only. For such methods (e.g. the implicit trapezoidal rule) S-stability is equivalent to A-stability while for other methods (e.g. the implicit midpoint rule) S-stability is more restrictive. (This "discrimination" against certain methods is also responsible for some of the unexpected results in [8].)

We would rather use the differential equation

(4.3) $$z' = \frac{\lambda(t)}{\varepsilon} z + \frac{1}{\varepsilon} g(t)$$

as our simplest model; the essential generalization over (4.1) is the admission of variable λ. Here $\tilde{z}(t)$ is normally not representable in closed form; for smooth g and λ we have

$$\tilde{z}(t) = - \frac{g(t)}{\lambda(t)} + O(\varepsilon).$$

If we choose $\lambda(t)$ arbitrarily and $g(t) = -\lambda(t) \exp(-\int_0^t \lambda(\tau)d\tau)$ then

$$\tilde{z}(t) = - \frac{1}{1+\varepsilon} \frac{g(t)}{\lambda(t)}.$$

Some special cases which form useful test examples are

$$\lambda(t) = -1, \qquad g(t) = e^t \qquad \tilde{z}(t) = \frac{1}{1+\varepsilon} e^t$$

$$\lambda(t) = -\frac{1}{1+t}, \qquad g(t) = 1 \qquad \tilde{z}(t) = \frac{1+t}{1+\varepsilon}$$

$$\lambda(t) = -e^{-t}, \qquad g(t) = \exp(1-t-e^{-t}) \qquad \tilde{z}(t) = \frac{1}{1+\varepsilon} \exp(1-e^{-t})$$

With (4.3) it is not difficult to analyze the behavior of important classes of implicit RK-methods (compare the analogous investigations for the model (4.1) in [8]).

Let a RK-method be characterized by its coefficient scheme $\begin{pmatrix} A & \vdots & c \\ \cdots & \vdots & \cdots \\ b^T & \vdots & \end{pmatrix}$.

Consider one step of the discretization of (4.3) leading from the value ζ_{n-1} at t_{n-1} to ζ_n at $t_n = t_{n-1} + h$ see, e.g., [1], section 3.1).

$$\zeta_n = \left[1 + \frac{h}{\varepsilon} b^T \Lambda_n \left(I - \frac{h}{\varepsilon} A\Lambda_n\right)^{-1} e\right] \zeta_{n-1}$$
$$+ \frac{h}{\varepsilon} b^T \left[I + \frac{h}{\varepsilon} \Lambda_n \left(I - \frac{h}{\varepsilon} A\Lambda_n\right)^{-1} A\right] g_n,$$

here $\Lambda_n := \text{diag}\,(\lambda(t_{n-1} + c_\mu h),\ \mu = 1(1)m)$, $e := (1,1,\ldots,1)^T$, $g_n := (g(t_{n-1} + c_\mu h),\ \mu = 1(1)m)^T$. Assume that $\lambda(t) \neq 0$. Then

$$\zeta_n = \left[1 + b^T \left(\frac{\varepsilon}{h} \Lambda_n^{-1} - A\right)^{-1} e\right]\zeta_{n-1} + \frac{h}{\varepsilon} b^T \left[I + \left(\frac{\varepsilon}{h} \Lambda_n^{-1} - A\right)^{-1} A\right] g_n.$$

Obviously the analysis for $\varepsilon \to 0$ is considerably simplified if A is nonsingular and its spectrum is in the right half-plane; then the inverse matrices $\left(\frac{\varepsilon}{h} \Lambda_n^{-1} - A\right)^{-1}$ exist for arbitrary values of $\frac{\varepsilon}{h} \geq 0$ and their limit for $\varepsilon \to 0$ is $-A^{-1}$ <u>independently of the variability of λ</u>. E.g., the "growth factor" per step tends to $1 - b^T A^{-1} e$ independently of $\lambda(t)$.

If A is singular the limit value of this growth factor (if it exists at all) will normally depend on the ratio of the $\lambda(t_{n-1} + \mu h)$:

<u>Example</u>: Exponentially fitted trapezoidal rule, $\frac{1}{2} < \gamma < 1$:

$$\left(\begin{array}{cc|c} 0 & 0 & 0 \\ 1-\gamma & \gamma & 1 \\ \hline 1-\gamma & \gamma & \end{array}\right) \qquad \frac{\varepsilon}{h} \Lambda_n^{-1} - A = \begin{pmatrix} \frac{\varepsilon}{h\lambda_{n-1}} & 0 \\ \gamma-1 & \frac{\varepsilon}{h\lambda_n} - \gamma \end{pmatrix}$$

$$1 + b^T \left(\frac{\varepsilon}{h} \Lambda_n^{-1} - A\right)^{-1} e = 1 + h\, \frac{(1-\gamma)\lambda_{n-1} + \gamma\lambda_n}{\varepsilon - h\lambda_n\gamma} \to -\frac{1-\gamma}{\gamma} \frac{\lambda_{n-1}}{\lambda_n} .$$

For the implicit trapezoidal rule proper ($\gamma = \frac{1}{2}$), the limit value $-\lambda_{n-1}/\lambda_n$ of the growth factor is greater than 1 in modulus if $\lambda(t)$ decreases in modulus. For $\gamma > \frac{1}{2}$, one gains a safety margin to allow for variable λ.

With <u>multistep methods</u>, the boundedness of the solution of discretizations of (4.3) is quite difficult to establish for general variable λ. The more important are the results on a certain class of multistep methods which have been obtained by Dahlquist ([7]).

It is hoped that rather comprehensive results can be compiled on the behavior of many types of discretization methods for the model equation (4.3) and that these results permit some preliminary assertions on the behavior of these methods. However, (4.3) does not permit the study of effects related to

a) decomposition consistency;
b) coupling between regular and singular components.

A more general but still feasible model system is the following:

$$(4.4)\quad \begin{aligned} x' &= \frac{1}{\varepsilon} a_{11}(t)\, x + \frac{1}{\varepsilon} a_{12}(t)\, y + \frac{1}{\varepsilon} g_1(t), \qquad a_{11}(t) < 0, \\ y' &= a_{21}(t)\, x + g_2(t). \end{aligned}$$

For given initial values x_o, y_o, the regular component of the solution of (4.4) has the initial values

$$\begin{aligned} \tilde{x}(0,\varepsilon) &= -\frac{1}{a_{11}(0)} (a_{12}(0) y_o + g_1(0)) + O(\varepsilon), \\ \tilde{y}(0,\varepsilon) &= y_o + O(\varepsilon). \end{aligned}$$

Thus the regular component of the solution of a discretization of (4.4) should have the same initial values except for terms of order $O(h^{q_o})$ and $O(\varepsilon)$. The implicit Euler method is decomposition consistent for (4.4), with $q_o = 1$.

V. CONCLUSIONS

Although (4.4) is still a rather special case of (S), it seems that an understanding of the behavior of discretizations of (4.4) will constitute a crucial step in the analysis of discretizations of general systems (S). The result for (4.4) should carry over at least to systems (S) where the $\frac{1}{\varepsilon}$ factor affects only linear terms in f.

It is hoped that it may be possible to design a framework of reference concepts and results on the basis of the ideas put forward in this paper. Such a "reference system" would permit a clearer assessment of the many contributions to the numerical treatment of stiff systems which have appeared within the last few years. I hope that this will lead, in due time, to a theory for discretizations of stiff systems which will prove as useful for the construction of sound numerical algorithms for such problems as the Dahlquist-Henrici theory of "$h \to 0$" has proved for the successful numerical solution of non-stiff systems of ordinary differential equations.

REFERENCES:

[1] H.J. STETTER: Analysis of discretization methods for ordinary differential equations, Springer-Verlag, 1973.

[2] G. DAHLQUIST: The sets of smooth solutions of differential and difference equations, in Stiff Differential Systems (ed. R. Willoughby), Plenum Press, 1974.

[3] W.L. MIRANKER: Numerical methods of boundary layer type for stiff systems of differential equations, Computing 11 (1973) 221-234.

[4] M. van VELDHUIZEN: Convergence of one-step discretizations for stiff differential equations, Thesis, Univ. of Utrecht, 1973.

[5] W. WASOW: Asymptotic expansion for ordinary differential equations, Interscience, 1965.

[6] C.W. GEAR: Numerical initial value problems in ordinary differential equations, Prentice Hall, 1971.

[7] G. DAHLQUIST: Error Analysis for a class of methods for stiff non-linear initial value problems, (these Proceedings).

[8] A. PROTHERO and A. ROBINSON: On the stability and accuracy of one-step methods for solving stiff systems of ordinary differential equations, Math. Comp. 28 (1974), 145-162.

Vol. 342: Algebraic K-Theory II, "Classical" Algebraic K-Theory, and Connections with Arithmetic. Edited by H. Bass. XV, 527 pages. 1973.

Vol. 343: Algebraic K-Theory III, Hermitian K-Theory and Geometric Applications. Edited by H. Bass. XV, 572 pages. 1973.

Vol. 344: A. S. Troelstra (Editor), Metamathematical Investigation of Intuitionistic Arithmetic and Analysis. XVII, 485 pages. 1973.

Vol. 345: Proceedings of a Conference on Operator Theory. Edited by P. A. Fillmore. VI, 228 pages. 1973.

Vol. 346: Fučík et al., Spectral Analysis of Nonlinear Operators. II, 287 pages. 1973.

Vol. 347: J. M. Boardman and R. M. Vogt, Homotopy Invariant Algebraic Structures on Topological Spaces. X, 257 pages. 1973.

Vol. 348: A. M. Mathai and R. K. Saxena, Generalized Hypergeometric Functions with Applications in Statistics and Physical Sciences. VII, 314 pages. 1973.

Vol. 349: Modular Functions of One Variable II. Edited by W. Kuyk and P. Deligne. V, 598 pages. 1973.

Vol. 350: Modular Functions of One Variable III. Edited by W. Kuyk and J.-P. Serre. V, 350 pages. 1973.

Vol. 351: H. Tachikawa, Quasi-Frobenius Rings and Generalizations. XI, 172 pages. 1973.

Vol. 352: J. D. Fay, Theta Functions on Riemann Surfaces. V, 137 pages. 1973.

Vol. 353: Proceedings of the Conference on Orders, Group Rings and Related Topics. Organized by J. S. Hsia, M. L. Madan and T. G. Ralley. X, 224 pages. 1973.

Vol. 354: K. J. Devlin, Aspects of Constructibility. XII, 240 pages. 1973.

Vol. 355: M. Sion, A Theory of Semigroup Valued Measures. V, 140 pages. 1973.

Vol. 356: W. L. J. van der Kallen, Infinitesimally Central Extensions of Chevalley Groups. VII, 147 pages. 1973.

Vol. 357: W. Borho, P. Gabriel und R. Rentschler, Primideale in Einhüllenden auflösbarer Lie-Algebren. V, 182 Seiten. 1973.

Vol. 358: F. L. Williams, Tensor Products of Principal Series Representations. VI, 132 pages. 1973.

Vol. 359: U. Stammbach, Homology in Group Theory. VIII, 183 pages. 1973.

Vol. 360: W. J. Padgett and R. L. Taylor, Laws of Large Numbers for Normed Linear Spaces and Certain Fréchet Spaces. VI, 111 pages. 1973.

Vol. 361: J. W. Schutz, Foundations of Special Relativity: Kinematic Axioms for Minkowski Space-Time. XX, 314 pages. 1973.

Vol. 362: Proceedings of the Conference on Numerical Solution of Ordinary Differential Equations. Edited by D.G. Bettis. VIII, 490 pages. 1974.

Vol. 363: Conference on the Numerical Solution of Differential Equations. Edited by G. A. Watson. IX, 221 pages. 1974.

Vol. 364: Proceedings on Infinite Dimensional Holomorphy. Edited by T. L. Hayden and T. J. Suffridge. VII, 212 pages. 1974.

Vol. 365: R. P. Gilbert, Constructive Methods for Elliptic Equations. VII, 397 pages. 1974.

Vol. 366: R. Steinberg, Conjugacy Classes in Algebraic Groups (Notes by V. V. Deodhar). VI, 159 pages. 1974.

Vol. 367: K. Langmann und W. Lütkebohmert, Cousinverteilungen und Fortsetzungssätze. VI, 151 Seiten. 1974.

Vol. 368: R. J. Milgram, Unstable Homotopy from the Stable Point of View. V, 109 pages. 1974.

Vol. 369: Victoria Symposium on Nonstandard Analysis. Edited by A. Hurd and P. Loeb. XVIII, 339 pages. 1974.

Vol. 370: B. Mazur and W. Messing, Universal Extensions and One Dimensional Crystalline Cohomology. VII, 134 pages. 1974.

Vol. 371: V. Poenaru, Analyse Différentielle. V, 228 pages. 1974.

Vol. 372: Proceedings of the Second International Conference on the Theory of Groups 1973. Edited by M. F. Newman. VII, 740 pages. 1974.

Vol. 373: A. E. R. Woodcock and T. Poston, A Geometrical Study of the Elementary Catastrophes. V, 257 pages. 1974.

Vol. 374: S. Yamamuro, Differential Calculus in Topological Linear Spaces. IV, 179 pages. 1974.

Vol. 375: Topology Conference. Edited by R. F. Dickman Jr. and P. Fletcher. X, 283 pages 1974.

Vol. 376: I. J. Good and D. B. Osteyee, Information, Weight of Evidence. The Singularity between Probability Measures and Signal Detection. XI, 156 pages. 1974.

Vol. 377: A. M. Fink, Almost Periodic Differential Equations. VIII, 336 pages. 1974.

Vol. 378: TOPO 72 – General Topology and its Applications. Proceedings 1972. Edited by R. A. Alò, R. W. Heath and J. Nagata. XIV, 651 pages. 1974.

Vol. 379: A. Badrikian et S. Chevet, Mesures Cylindriques, Espaces de Wiener et Fonctions Aléatoires Gaussiennes. X, 383 pages. 1974.

Vol. 380: M. Petrich, Rings and Semigroups. VIII, 182 pages. 1974.

Vol. 381: Séminaire de Probabilités VIII. Edité par P. A. Meyer. IX, 354 pages. 1974.

Vol. 382: J. H. van Lint, Combinatorial Theory Seminar Eindhoven University of Technology. VI, 131 pages. 1974.

Vol. 383: Séminaire Bourbaki – vol. 1972/73. Exposés 418–435. IV, 334 pages. 1974.

Vol. 384: Functional Analysis and Applications, Proceedings 1972. Edited by L. Nachbin. V, 270 pages. 1974.

Vol. 385: J. Douglas Jr. and T. Dupont, Collocation Methods for Parabolic Equations in a Single Space Variable (Based on C^1-Piecewise-Polynomial Spaces). V, 147 pages. 1974.

Vol. 386: J. Tits, Buildings of Spherical Type and Finite BN-Pairs. X, 299 pages. 1974.

Vol. 387: C. P. Bruter, Eléments de la Théorie des Matroïdes. V, 138 pages. 1974.

Vol. 388: R. L. Lipsman, Group Representations. X, 166 pages. 1974.

Vol. 389: M.-A. Knus et M. Ojanguren, Théorie de la Descente et Algèbres d' Azumaya. IV, 163 pages. 1974.

Vol. 390: P. A. Meyer, P. Priouret et F. Spitzer, Ecole d'Eté de Probabilités de Saint-Flour III – 1973. Edité par A. Badrikian et P.-L. Hennequin. VIII, 189 pages. 1974.

Vol. 391: J. W. Gray, Formal Category Theory: Adjointness for 2-Categories. XII, 282 pages. 1974.

Vol. 392: Géométrie Différentielle, Colloque, Santiago de Compostela, Espagne 1972. Edité par E. Vidal. VI, 225 pages. 1974.

Vol. 393: G. Wassermann, Stability of Unfoldings. IX, 164 pages. 1974.

Vol. 394: W. M. Patterson, 3rd, Iterative Methods for the Solution of a Linear Operator Equation in Hilbert Space – A Survey. III, 183 pages. 1974.

Vol. 395: Numerische Behandlung nichtlinearer Integrodifferential- und Differentialgleichungen. Tagung 1973. Herausgegeben von R. Ansorge und W. Törnig. VII, 313 Seiten. 1974.

Vol. 396: K. H. Hofmann, M. Mislove and A. Stralka, The Pontryagin Duality of Compact O-Dimensional Semilattices and its Applications. XVI, 122 pages. 1974.

Vol. 397: T. Yamada, The Schur Subgroup of the Brauer Group. V, 159 pages. 1974.

Vol. 398: Théories de l'Information, Actes des Rencontres de Marseille-Luminy, 1973. Edité par J. Kampé de Fériet et C.-F. Picard. XII, 201 pages. 1974.

Vol. 399: Functional Analysis and its Applications. Proceedings 1973. Edited by H. G. Garnir, K. R. Unni and J. H. Williamson. II, 584 pages. 1974.

Vol. 400: A Crash Course on Kleinian Groups. Proceedings 1974. Edited by L. Bers and I. Kra. VII, 130 pages. 1974.

Vol. 401: M. F. Atiyah, Elliptic Operators and Compact Groups. V, 93 pages. 1974.

Vol. 402: M. Waldschmidt, Nombres Transcendants. VIII, 277 pages. 1974.

Vol. 403: Combinatorial Mathematics. Proceedings 1972. Edited by D. A. Holton. VIII, 148 pages. 1974.

Vol. 404: Théorie du Potentiel et Analyse Harmonique. Edité par J. Faraut. V, 245 pages. 1974.

Vol. 405: K. J. Devlin and H. Johnsbråten, The Souslin Problem. VIII, 132 pages. 1974.

Vol. 406: Graphs and Combinatorics. Proceedings 1973. Edited by R. A. Bari and F. Harary. VIII, 355 pages. 1974.

Vol. 407: P. Berthelot, Cohomologie Cristalline des Schémas de Caracteristique p > o. II, 604 pages. 1974.

Vol. 408: J. Wermer, Potential Theory. VIII, 146 pages. 1974.

Vol. 409: Fonctions de Plusieurs Variables Complexes, Séminaire François Norguet 1970–1973. XIII, 612 pages. 1974.

Vol. 410: Séminaire Pierre Lelong (Analyse) Année 1972–1973. VI, 181 pages. 1974.

Vol. 411: Hypergraph Seminar. Ohio State University, 1972. Edited by C. Berge and D. Ray-Chaudhuri. IX, 287 pages. 1974.

Vol. 412: Classification of Algebraic Varieties and Compact Complex Manifolds. Proceedings 1974. Edited by H. Popp. V, 333 pages. 1974.

Vol. 413: M. Bruneau, Variation Totale d'une Fonction. XIV, 332 pages. 1974.

Vol. 414: T. Kambayashi, M. Miyanishi and M. Takeuchi, Unipotent Algebraic Groups. VI, 165 pages. 1974.

Vol. 415: Ordinary and Partial Differential Equations. Proceedings 1974. XVII, 447 pages. 1974.

Vol. 416: M. E. Taylor, Pseudo Differential Operators. IV, 155 pages. 1974.

Vol. 417: H. H. Keller, Differential Calculus in Locally Convex Spaces. XVI, 131 pages. 1974.

Vol. 418: Localization in Group Theory and Homotopy Theory and Related Topics. Battelle Seattle 1974 Seminar. Edited by P. J. Hilton. VI, 172 pages 1974.

Vol. 419: Topics in Analysis. Proceedings 1970. Edited by O. E. Lehto, I. S. Louhivaara, and R. H. Nevanlinna. XIII, 392 pages. 1974.

Vol. 420: Category Seminar. Proceedings 1972/73. Edited by G. M. Kelly. VI, 375 pages. 1974.

Vol. 421: V. Poénaru, Groupes Discrets. VI, 216 pages. 1974.

Vol. 422: J.-M. Lemaire, Algèbres Connexes et Homologie des Espaces de Lacets. XIV, 133 pages. 1974.

Vol. 423: S. S. Abhyankar and A. M. Sathaye, Geometric Theory of Algebraic Space Curves. XIV, 302 pages. 1974.

Vol. 424: L. Weiss and J. Wolfowitz, Maximum Probability Estimators and Related Topics. V, 106 pages. 1974.

Vol. 425: P. R. Chernoff and J. E. Marsden, Properties of Infinite Dimensional Hamiltonian Systems. IV, 160 pages. 1974.

Vol. 426: M. L. Silverstein, Symmetric Markov Processes. X, 287 pages. 1974.

Vol. 427: H. Omori, Infinite Dimensional Lie Transformation Groups. XII, 149 pages. 1974.

Vol. 428: Algebraic and Geometrical Methods in Topology, Proceedings 1973. Edited by L. F. McAuley. XI, 280 pages. 1974.

Vol. 429: L. Cohn, Analytic Theory of the Harish-Chandra C-Function. III, 154 pages. 1974.

Vol. 430: Constructive and Computational Methods for Differential and Integral Equations. Proceedings 1974. Edited by D. L. Colton and R. P. Gilbert. VII, 476 pages. 1974.

Vol. 431: Séminaire Bourbaki – vol. 1973/74. Exposés 436–452. IV, 347 pages. 1975.

Vol. 432: R. P. Pflug, Holomorphiegebiete, pseudokonvexe Gebiete und das Levi-Problem. VI, 210 Seiten. 1975.

Vol. 433: W. G. Faris, Self-Adjoint Operators. VII, 115 pages. 1975.

Vol. 434: P. Brenner, V. Thomée, and L. B. Wahlbin, Besov Spaces and Applications to Difference Methods for Initial Value Problems. II, 154 pages. 1975.

Vol. 435: C. F. Dunkl and D. E. Ramirez, Representations of Commutative Semitopological Semigroups. VI, 181 pages. 1975.

Vol. 436: L. Auslander and R. Tolimieri, Abelian Harmonic Analysis, Theta Functions and Function Algebras on a Nilmanifold. V, 99 pages. 1975.

Vol. 437: D. W. Masser, Elliptic Functions and Transcendence. XIV, 143 pages. 1975.

Vol. 438: Geometric Topology. Proceedings 1974. Edited by L. C. Glaser and T. B. Rushing. X, 459 pages. 1975.

Vol. 439: K. Ueno, Classification Theory of Algebraic Varieties and Compact Complex Spaces. XIX, 278 pages. 1975

Vol. 440: R. K. Getoor, Markov Processes: Ray Processes and Right Processes. V, 118 pages. 1975.

Vol. 441: N. Jacobson, PI-Algebras. An Introduction. V, 115 pages. 1975.

Vol. 442: C. H. Wilcox, Scattering Theory for the d'Alembert Equation in Exterior Domains. III, 184 pages. 1975.

Vol. 443: M. Lazard, Commutative Formal Groups. II, 236 pages. 1975.

Vol. 444: F. van Oystaeyen, Prime Spectra in Non-Commutative Algebra. V, 128 pages. 1975.

Vol. 445: Model Theory and Topoi. Edited by F. W. Lawvere, C. Maurer, and G. C. Wraith. III, 354 pages. 1975.

Vol. 446: Partial Differential Equations and Related Topics. Proceedings 1974. Edited by J. A. Goldstein. IV, 389 pages. 1975.

Vol. 447: S. Toledo, Tableau Systems for First Order Number Theory and Certain Higher Order Theories. III, 339 pages. 1975.

Vol. 448: Spectral Theory and Differential Equations. Proceedings 1974. Edited by W. N. Everitt. XII, 321 pages. 1975.

Vol. 449: Hyperfunctions and Theoretical Physics. Proceedings 1973. Edited by F. Pham. IV, 218 pages. 1975.

Vol. 450: Algebra and Logic. Proceedings 1974. Edited by J. N. Crossley. VIII, 307 pages. 1975.

Vol. 451: Probabilistic Methods in Differential Equations. Proceedings 1974. Edited by M. A. Pinsky. VII, 162 pages. 1975.

Vol. 452: Combinatorial Mathematics III. Proceedings 1974. Edited by Anne Penfold Street and W. D. Wallis. IX, 233 pages. 1975.

Vol. 453: Logic Colloquium. Symposium on Logic Held at Boston, 1972–73. Edited by R. Parikh. IV, 251 pages. 1975.

Vol. 454: J. Hirschfeld and W. H. Wheeler, Forcing, Arithmetic, Division Rings. VII, 266 pages. 1975.

Vol. 455: H. Kraft, Kommutative algebraische Gruppen und Ringe. III, 163 Seiten. 1975.

Vol. 456: R. M. Fossum, P. A. Griffith, and I. Reiten, Trivial Extensions of Abelian Categories. Homological Algebra of Trivial Extensions of Abelian Categories with Applications to Ring Theory. XI, 122 pages. 1975.